ANCIENT SEDIMENTARY ENVIRONMENTS

ANCIENT SEDIMENTARY ENVIRONMENTS

A Brief Survey

RICHARD C. SELLEY

CHAPMAN AND HALL LTD
LONDON

First published 1970
© 1970 *Richard C. Selley*
Reprinted 1972
Reprinted 1973
Set in Photon Times 11 on 12 pt. by
Richard Clay (The Chaucer Press) Ltd., Bungay, Suffolk
and printed in Great Britain by
Fletcher & Son Ltd, Norwich, Norfolk
S B N 412 10100 9

The geologist of the 1970s will be a great working combination of geophysicist, geologist, computer-man, mapmaker, and businessman. But the good old rockhound type will abound and be in more demand than ever before. As the search for stratigraphic oil traps becomes serious, the geologist will be and is now going back to the field like never before. More emphasis is being placed on paleontology, geomorphology, paleogeography, paleogeomorphology, and other related earth studies. Ancient deltas, shorelines, bars, and stream beds take on new meaning as the search for stratigraphic oil traps moves out.

J. C. MCCASLIN
'Forecast for the Seventies'
Oil & Gas Journal 10 November 1969.

CONTENTS

PREFACE

In this book I have attempted to show how the depositional environments of sedimentary rocks can be recognized. This is not a work for the specialist sedimentologist, but an introductory survey for readers with a basic knowledge of geology.

Within the last few years environmental analysis of ancient sediments has been enhanced by intensive studies of their modern counterparts. Thus Geikie's dictum 'the present is the key to the past' can now be applied with increasing accuracy. While an understanding of Recent processes and environments is critical to the interpretation of their ancient analogues, it is beyond the scope of this book to describe these studies in detail. I have, however, attempted to summarize the results of this work; detail being sacrificed in the interest of brevity. Inevitably, this has led to a tendency to generalize; I have tried to counteract this by providing bibliographies of studies describing Recent sediments.

The economic importance of environmental analysis of ancient sediments is increasing. With the world expected to consume as much oil and gas in the decade 1970–80 as has been produced to date, the search extends increasingly to the more elusive stratigraphically controlled accumulations.

Similarly environmental analysis has a part to play in locating metallic ores in sediments whose geometries are facies controlled.

The book begins with a discussion of the classification of sedimentary environments and an evaluation of the methods which may be used to identify them in ancient deposits.

Each subsequent chapter describes a particular depositional environment, beginning with a summary of its characteristics as seen on the earth's surface at the present time. This is followed by a description of an ancient case history whose origin is then deduced. A general discussion of the problems of identifying the environment in ancient sediments comes next, and each chapter concludes with a brief review of its economic significance.

Neither the selection of environments nor the discussion of economic aspects is intended to be comprehensive. I hope, however, that there are sufficient examples to show how the environment of a sedimentary rock can be determined and some of the ways in which sedimentology can be used in the search for economic materials. The discussions of the economic aspects of the various environments are heavily biased in favour of the oil industry at the expense of mining, hydrogeology, and engineering geology. This is no accident. The oil industry is the largest employer of sedimentologically oriented geologists and has done more to advance and apply sedimentology than any other branch of industrial geology.

Critical readers will notice that metres, feet, kilometres, and miles are used indiscriminately throughout the book. Since the oil industry refuses to go metric, the student must quickly learn to correlate the two systems. A conversion scale is included in the first figure.

January, 1970   RICHARD C. SELLEY
Tripoli, Libya

ACKNOWLEDGEMENTS

Any textbook, by its very nature, is parasitic on previously published work. This is probably more true of this book than many others due to its case history approach.

I am, therefore, extremely grateful to the authors of the various case histories who allowed their work to be pirated in this fashion, who criticized portions of the manuscript, and who generously supplied photographs. These include J. R. L. Allen, A. H. Bouma, J. D. Collinson, A. Hallam, A. J. Jenik, J. F. Lerbekmo, N. D. Newell, M. D. Picard, H. G. Reading, D. J. Stanley, W. F. Tanner, G. S. Visher, R. G. Walker, and R. J. Weimer.

Correct opinions and environmental interpretations are attributable to the authors of the case histories. Errors of fact and interpretation are due to me.

My colleagues at the Oasis Geological Laboratory, particularly Dr J. Hea and D. Baird, suggested numerous improvements to the manuscript. Thanks are also due to Oasis Oil Company of Libya Inc. for permission to publish.

I am extremely grateful to Professor J. Sutton and the Imperial College authorities who granted me leave of absence to retire from the academic hurly-burly to the peace and quiet of the oil industry to write this book.

R. C. S.

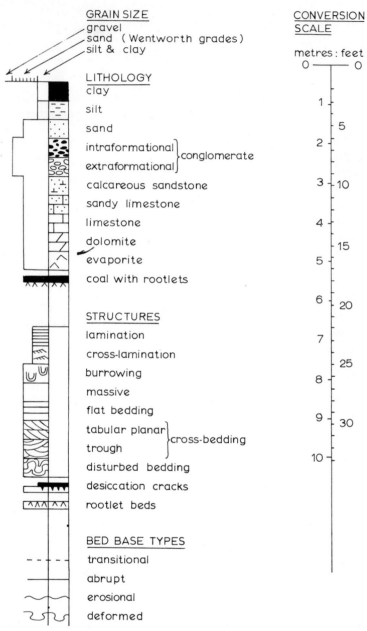

GRAIN SIZE
gravel
sand (Wentworth grades)
silt & clay

LITHOLOGY
clay
silt
sand
intraformational ⎫
                  ⎬ conglomerate
extraformational ⎭
calcareous sandstone
sandy limestone
limestone
dolomite
evaporite
coal with rootlets

STRUCTURES
lamination
cross-lamination
burrowing
massive
flat bedding
tabular planar ⎫
               ⎬ cross-bedding
trough   ⎭
disturbed bedding
desiccation cracks
rootlet beds

BED BASE TYPES
transitional
abrupt
erosional
deformed

CONVERSION SCALE

metres : feet
0 ——— 0
1
          5
2
3 — 10
4
          15
5
6
     20
7
          25
8
9   30
10

Figure 0.1. Detailed measured rock sections, illustrated throughout the book, are drawn using the above key.

CHAPTER ONE

INTRODUCTION

SEDIMENTARY ENVIRONMENTS AND FACIES

A *sedimentary environment* is a part of the earth's surface which is physically, chemically, and biologically distinct from adjacent terrains. Examples include deserts, river valleys, and deltas.

The three defining parameters listed above include the fauna and flora of the environment, its geology, geomorphology, climate, weather, and, if subaqueous, the depth, temperature, salinity, and current system of the water. These variables are tightly knit in dynamic equilibrium with one another like the threads of a spider's web. A change in one variable causes changes in all the others.

A sedimentary environment may be a site of erosion, non-deposition, or deposition. As a broad generalization, sub-aerial environments are typically erosional while sub-aqueous environments are mostly depositional areas. Some environments alternate through time between phases of erosion, equilibrium, and deposition. River valleys are a case in point.

Numerous deserts, lakes, deltas, reefs, and other environments are found all over the present day face of the earth suggesting that there are a finite number of sedimentary environments. This statement must be qualified though by noting that no two similar environments are ever exactly alike, and that different environments often merge imperceptibly with one another across the face of the earth.

A *sedimentary facies* is a mass of sedimentary rock which can be defined and distinguished from others by its geometry, lithology, sedimentary structures, palaeocurrent pattern, and fossils.

The consensus of geological opinion is that there are a finite number of sedimentary facies which occur repeatedly in rocks of different ages all over the world. Comparison with Recent sediments suggests that these can be related to present-day depositional environments. As with their Recent counterparts, therefore, no two similar sedimentary facies are ever identical, and gradational transitions between facies are common.

A sedimentary facies is the product of a depositional environment, a special kind of sedimentary environment. One of the main problems of determining the origin of ancient sediments is that, though essentially reflecting depositional environments, they also inherit features of earlier erosional and non-depositional phases. Consider for example an old river channel. The profile of the stream bed reflects an erosional environment, the infilling sediment reflects the nature of the source rocks and the hydraulics of the current which transported (as well as deposited) the sediment, while rolled bones and wood are derived from non-depositional environments outside the channel. It is only the sedimentary structures (and palaeocurrents) which unequivocally indicate the depositional environment.

These concepts of sedimentary environments and facies are summarized in Table 1.1.

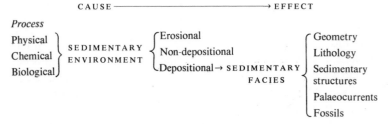

Table 1.1. The relationship between sedimentary environments and sedimentary facies.

It is very important to distinguish a sedimentary environment from a sedimentary facies. There is no problem in identifying the environment of Recent sediments. A sand sample collected from a beach today is, by definition, a beach sand. When studying ancient sediments, however, it is best to classify them into facies on a purely descriptive basis; it is unwise to give them environmental names. Thus one should talk of pebbly channel sand facies, flysch facies, and so on, rather than of fluviatile facies or turbidite facies.

Many attempts have been made to classify both Recent sedimentary environments and ancient sedimentary facies (e.g. Pettijohn, 1957, p. 610; Krumbein and Sloss, 1959, p. 196; Visher, 1965; Potter, 1967 and Shelton, 1967).

The classification of depositional environments used in this book is unique, but owes much to previous ones (Table 1.2). It is put forward not as a final statement of environments but as a foundation on which to base the case histories discussed in subsequent chapters.

Recent sedimentary environments can generally be divided into sub-environments. For example, a linear shoreline is often composed of a complex of barrier islands, lagoons, and tidal flats lying between

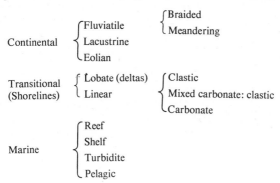

Table 1.2. A classification of depositional environments. Like all classifications this one contains several deficiencies and inconsistencies which are inherent due to the complexity of environments. Note particularly that eolian deposits can form on the crests of barrier islands: deltas can build into lakes as well as seas; reefs occur in fresh, as well as sea water. Strictly speaking, a turbidite is a process rather than an environment which, though typically marine, can occur in many situations.

alluvial and shelf major environments. Ancient sedimentary facies can be divided in a similar manner into sub-facies which can often be attributed to sub-environments.

METHODS OF ENVIRONMENTAL DIAGNOSIS

There are many different techniques which can be used to determine the depositional environment of a sedimentary rock. An alternative arrangement for this book would have been to describe and discuss methods of environmental recognition, rather than to actually show them in operation solving specific case histories. Though the latter approach seems more interesting it is appropriate first to discuss the various techniques which will shortly be employed.

As a general rule the diagnosis of a sedimentary environment should be based on a critical evaluation of all available lines of evidence. Commonly however not all the data that are desirable are actually available. This is particularly true of sub-surface studies based on well logs. In such cases one can only make the best use of what facts can be ascertained. In other cases, such as outcrop studies, though there may be a wealth of facts to utilize, only one or two may actually be critical to the diagnosis of the environment in question. In different cases different lines of evidence will be of varying importance. Thus the recognition of a reef may be confidently based on only lithological and palaeontological data. A delta on the other hand may be identified by its geometry and vertical sequence of grain size and sedimentary structures. These points will become clearer as the book proceeds.

The techniques of environmental analysis can most conveniently be discussed under the five defining parameters of a facies: geometry, lithology, sedimentary structures, palaeocurrent patterns, and fossils.

Geometry

The over-all shape of a sedimentary facies is a function of pre-depositional topography, the geomorphology of the depositional environment and its post-depositional history. For example the geometry of blanket sands burying old land surfaces will largely be controlled by the geomorphic stage and process which previously operated on that terrain. This may give little indication of the origin of the overlying sand. Likewise post-depositional erosion and tectonic deformation will modify the geometry of a facies and inhibit the use of its over-all shape as a diagnostic feature. It is only when the syn-depositional shape of an environment is preserved that geometry is a valuable diagnostic criterion. The radiating channel patterns of deltaic distributaries, and the shoestring shapes of offshore bars are examples of this.

The geometry of a sedimentary facies may be relatively easy to determine where it crops out at the surface and where exposure is good. Unfortunately it is in those cases where geometry is most important, as in the location of oil reservoirs, that it is hardest to determine. In such situations other techniques must first be used to identify the environment. This knowledge can then be used to predict the geometry of the reservoir-bearing facies.

Lithology

The lithology of a sedimentary facies is one of the easiest of parameters to observe and one of considerable environmental significance. As a general rule the lithology of limestones is of more diagnostic importance than that of sandstones. This is because studies of Recent carbonates show that they occur in a number of micro-facies whose distribution is closely related to their depositional environment (e.g. Imbrie and Purdy, 1962). Furthermore, they cannot survive transportation far from where they formed (Ginsburg and others, 1963, p. 572). Because of these facts many limestones can be attributed to an environment by examination of a small chip or thin section and close comparison with Recent deposits (see Chapters 8 and 9).

The lithology of a clastic sediment on the other hand is a function not only of the environment in which it was deposited but also of its transportational history and of the type of rock from which it was derived. Petrographic studies of sandstones are thus of much less value as an indicator of depositional environment than those of carbonates. On the other hand sandstones are less susceptible to diagenesis than carbonates, so their depositional fabric is generally easier to discern.

Many attempts have been made to use the texture of a sediment to determine its depositional environment. This topic is described in a vast literature which can scarcely be avoided by leafing through any recent textbook or journal dealing with sedimentary rocks. Folk (1966) gives a recent review of the status of this work. The basic philosophy of this technique is that the sediments of Recent environments can be distinguished from one another by careful statistical evaluation of their texture. By analogy it should be possible to carry out granulometric analyses of ancient sediments, subject them to various statistical gymnastics and, by comparing them with Recent sediments of known origin, determine their depositional environment.

Unfortunately this approach has largely proved unsatisfactory and the granulometric analysis of ancient sediments is a declining art, to the delight of laboratory technicians. There are several reasons for the failure of this method. First, as already mentioned, the texture of a sediment is a function not just of the depositional environment, but also of its previous history. Thus, to take an extreme case, if a clean well-sorted fine sand is the only sediment being transported into an environment then a clean well-sorted fine sand is all that can be deposited in it. A second problem of using texture to determine the environment of a sediment lies in the origin of the fine detritus. Studies of Recent sediments show that clay content is a very sensitive indicator of depositional process (Doeglas, 1946). In an ancient sediment it is not possible to prove that the clay matrix was deposited at the same time as the rest of the detritus. It is also possible that it was washed in at a later date, that it was transported in particles coarser than clay, or that it was formed from the diagenetic break-down of chemically unstable sand grains (Cummings, 1962). A third problem of the granulometric analysis of sedimentary rocks is a purely technical one. Post-depositional solution and overgrowths may considerably alter the size of sand grains and modify the over-all texture of a rock. Intensive cementation may make it impossible for a rock to be disaggregated to its original textural composition. In such cases it may be necessary to determine the sorting curve by measuring the dimensions of grains in thin sections. This is a slow and complex process and the end results are extremely difficult to compare with sieve analyses of Recent sediment samples (Chayes, 1956).

For the reasons outlined above, therefore, statistical textural studies of ancient sediments have largely proved an unsatisfactory method of environmental diagnosis.

Considered from a more general standpoint the grain size of a sediment is an important indicator of the energy level of a depositional environment. The coarser the grain size the higher the energy level of the depositing current and the better the sorting the more prolonged its action. These generalizations have long been applied to terrigenous sediments and form the basis for a useful classification of carbonates (Dunham, 1962). However even these generalizations of the correlation of grain size and sorting to energy level must be interpreted with caution. No matter how strong a current it cannot deposit sediment coarser than that available in the source. Similarly the grain size of carbonate rocks may not always be a reliable indicator of the energy level of their depositional environment. This is because many limestones are composed of skeletal particles. Shells can form what are virtually organic conglomerates in a low-energy environment merely by living and dying there. Lagoonal oyster reefs are a case in point. Similarly the sorting and micrite content of carbonates is not solely a function of turbulence but is also influenced by other factors, notably micrite-forming algae and shell-munching predators (Bathurst, 1967, pp. 463–4).

Considerable attention has been given to the way in which depositional processes may affect the shape of sand grains. Recent glacial sands tend to be more angular and of lower sphericity than water-borne grains, while dune sands are often very well rounded. Kuenen (1960) summarizes a series of papers describing the experimental abrasion of sand grains by various processes. His data confirm that wind action is a more effective rounding agent than running water. Grain shape however is not just a function of the last depositional process which affected it but also of its previous history and original shape. Polycyclic sands will tend to be well-rounded regardless of the various processes to which they have been subjected (see also p. 60).

Electron microscope examination has shown that glacial, aqueous and eolian processes imprint characteristic markings on grain surfaces (e.g. Krinsley and Funnel, 1965). This is a valuable technique but it must be interpreted with care. Post-depositional solution, compaction, and tectonic grinding may modify the surface texture of a sand grain. It may also take a while for a grain which has been transported from one environment to another to have its surface altered.

Apart from its physical aspects the chemistry of a sediment holds many important clues as to its origin. Degens (1965) gives a comprehensive account of this topic. Only one or two examples will be described here to illustrate the present scope and future potential of geochemistry in environmental diagnosis.

There has been considerable discussion on the use of clay minerals as indicators of sedimentary environments (e.g. Grim, 1958). As a general rule, however, the chemistry of a clay reflects not just its depositional environment but also the parent rock, the climate which weathered it, and its diagenetic history. Weaver (1958) on the basis of hundreds of analyses of clays from ancient sediments showed that no single clay mineral is specific to any one environment, even in such broad terms as marine and non-marine. His data do show however that illite and montmorillonite are more characteristic of ancient marine rocks, while kaolin is more typical of continental, especially fluviatile deposits (ibid., p. 258, Fig. 1).

Glauconite is a mineral which is widely believed to form only in marine environments and to be so unstable that it cannot survive reworking. It is therefore held to be diagnostic of marine rocks. This criterion is not infallible. Detrital glauconite occurs in fluviatile Neogene red beds in the Dead Sea Valley. As a general rule however glauconite is a useful indicator of marine environments and studies of the present day distribution of glauconite suggest that it may be possible to narrow down the precise conditions which limit its formation (Porrenga, 1967).

The Boron content of clays was first discussed as a palaeosalinity indicator by Degens and others in 1957, and has subsequently been studied by many workers. Analysis of 82 Recent muds has shown that, for a given clay content, marine muds contain 30–45 ppm more Boron than freshwater muds (Shimp and others, 1969). This technique for distinguishing non-marine and marine deposits has been applied to ancient rocks with varying degrees of success (see for example Potter and others, 1963).

Similar geochemical approaches have been applied to other constituents of Recent sediments, notably phosphates and iron, in attempts to diagnose their depositional environments in ancient rocks. These studies have not so far been notably successful due largely to the scarcity of Recent environments where phosphates and iron are forming (see for example Bromley, 1967 and Rohrlich and others, 1969, respectively).

In conclusion it can be seen that the lithology of a sedimentary facies holds many important clues to its depositional environment. This is truer of carbonates than sandstones since they are deposited at, or close by, their point of origin. The lithology of sandstones gives less indication of their depositional environment since the sediment is introduced from outside the site of deposition and inherits extraneous characteristics due to its previous history.

Grain size, sorting, shape, and texture often reflect the energy level and process of the environment. They must be interpreted carefully. Geochemistry, a rapidly expanding field, has great potential in environmental diagnosis.

Sedimentary structures

Sedimentary structures are very important indicators of depositional environment. Unlike lithology and fossils they are undoubtedly generated in place and can never have been brought in from outside.

Sedimentary structures are easy to study at outcrop where exposure is good. In sub-surface work, however, only the smallest ones can occasionally be found in rock cores.

A huge number of sedimentary structures have been described in the literature. Their nomenclature and classification is confused and complex. This is largely because it is extremely hard to define accurately their morphology. Atlases of sedimentary structures have, however, been compiled by Pettijohn and Potter (1964), Gubler and others (1966), and Conybeare and Crook (1968).

Within recent years the interpretation of sedimentary structures has been greatly enhanced both by experimental work and by studies of Recent environments.

Sedimentary structures can provide evidence of whether an environment was glacial, aqueous, or sub-aerial. They give some indication of the depth and energy level of the environment (Allen, 1967) and of the velocity, hydraulics, and direction of the currents which flowed across it.

Most sedimentary structures can be arbitrarily fitted into a genetic classification of pre- syn- and post-depositional categories. The environmental significance of sedimentary structures will now be briefly reviewed under these three headings.

Pre-depositional sedimentary structures
Pre-depositional sedimentary structures are those observed on bed interfaces which formed before deposition of the younger bed. They are thus erosional features, not to be confused with post-depositional bed base deformational phenomena such as loadcasts. Pre-depositional structures include channels, scour marks, flutes, grooves, tool markings, and a host of other erosional phenomena. Despite considerable experimental work the hydraulic conditions which cause these structures are not well understood. Regardless of this they give valuable indications of the flow directions of the current which generated them.

Syn-depositional sedimentary structures
Structures formed during deposition include flat-bedding, cross-bedding, lamination, and micro-crosslamination (ripple-marking).

When a bed of sand is subjected to a current of increasing velocity different bed forms develop in a regular sequence. Each bed form deposits a different sedimentary structure (see Table 1.3). The point at which one bed form changes to another is a function of many variables, including the velocity, temperature, and viscosity of the fluid and the fall velocity (which more or less corresponds to the grain size) of the sediment particles. These variables have been integrated into the flow regime concept which expresses the sum total of all these parameters (Simons, Richardson, and Nordin, 1965). The

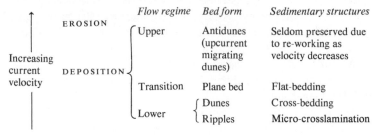

Table 1.3. The sequence of bed forms and sedimentary structures produced by a current of increasing velocity passing over a sand bed. The threshold velocity at which one bed form changes to another increases with increasing grain size. The ripple phase is absent for sands coarser than about 0·60 md. (Based on data in Simons, Richardson, and Nordin, 1965.)

main value of the flow regime concept is that it points out that beds of the same grain size need not necessarily have been deposited by currents of equal velocity. For example a micro-crosslaminated sand was deposited by a current of lower velocity than that which deposited a flat-bedded sand of equivalent grade. Similarly two beds of different grain size may not have been deposited by currents of different velocities. A flat-bedded fine sand and a cross-bedded medium sand may have been deposited by currents of the same velocity.

Despite the valuable insight that the flow regime concept has given to the study of syn-depositional sedimentary structures it has only indirectly aided environmental analysis. This is because the conditions generated by the confined flow of water in a sediment flume occur in nature in many different situations, such as rivers, estuaries, delta distributaries, tidal channels, and so on.

Cross-bedding is a sedimentary structure which is both common, morphologically variable, and extensively documented (see Potter and Pettijohn, 1963, pp. 62–89). Much can be learnt about the process which formed cross-bedding, but, since few processes are restricted to any one environment, this knowledge is of limited value (Allen, 1964). Large-scale eolian cross-bedding may be an exception to this rule (see p. 60).

Micro-crosslamination however is of more use as an environmental indicator. Walker (1963) and Jopling and Walker (1968) have described morphological criteria for distinguishing ripples due to different processes which are sometimes environmentally specific. Tanner (1967) has empirically determined a number of statistical criteria for identifying the depositional environment of ripples. Even so, Allen (1968) has shown how the most sophisticated of analytical and experimental techniques, while contributing much to our understanding of ripple formation, are of little value in environmental interpretation.

Post-depositional sedimentary structures
Structures which develop in sediments after their deposition are no less diverse and complex than those which form during deposition. They too have generated a ponderous literature in geological books and journals (see Potter and Pettijohn, 1963, Chapter 6 for a useful review). Again experimental studies, and the examination of Recent deposits have done much to increase our understanding of soft sediment deformation. Basically it is possible to distinguish two main genetic groups of these structures. One group shows an essentially vertical reorientation of bedding, while the second shows lateral rearrangement of the fabric. The first type includes loadcasts and pseudonodules where beds of sand collapse into underlying mud, as well as convolute lamination and quicksand structures which are contortions within beds of interlaminated fine muddy sand and clean sand respectively. Experimental studies have shown that there are a variety of triggering mechanisms which can generate these structures in unconsolidated sediment. These include earthquakes, current turbulences, and hydrostatic pressures due to connate fluids (see Selley, 1969a for a review). These processes can all operate in a variety of geological situations and are therefore of little significance in environmental analysis.

The second group of post-depositional sedimentary structures are those due to lateral movement of sediment, as opposed to the predominantly vertical reorientation shown by the previous type. These structures include slumps and slides which are penecontemporaneous recumbent folds and faults indicating large-scale sideways transport of sediment. Such phenomena occur where rapid sedimentation or erosion forms steep slopes which, from time to time, become so unstable that they collapse and shed superficial deposits down slope. These slope failures can occur spontaneously or be triggered by earthquakes, storms, or herds of stampeding dinosaurs.

Deformational sedimentary structures due to lateral movement can occur in many geological situations ranging from river banks to abyssal slopes. Slumping on an extensive scale however seems to be most typical of delta slopes and submarine canyons and fans.

Regardless of whether slumps can be used to determine depositional environment they are important criteria for recognizing palaeoslopes.

Palaeocurrent patterns

Of the five defining parameters of a sedimentary facies (geometry, lithology, structures, palaeocurrents, and fossils) all but palaeocurrents are observational criteria. To determine the palaeocurrents of a facies involves not just the description, but also the interpretation of data. This is a severe handicap. Compensation is to be found however in the fact that since palaeocurrents are discovered from sedimentary structures they clearly reflect the depositional environment of a facies and cannot inherit features from outside the actual site of deposition. The palaeocurrent analysis of a facies involves the following steps:

(*i*) Measurement of the orientation of significant sedimentary structures in the field (e.g. cross-bedding dip direction, channel axes, etc.).

(*ii*) Deduction of palaeocurrent direction at each sample point.

(*iii*) Preparation of regional palaeocurrent map.

(*iv*) Integration of palaeocurrent map with other lines of facies analysis to determine environment and palaeogeography. In some environments palaeocurrents may indicate palaeoslope (e.g. in rivers), in others they do not (e.g. eolian deposits).

The methodology and applications of palaeocurrent analysis have been reviewed by Potter and Pettijohn (1963). Many workers have warned that the deduction of palaeocurrent directions must be carried out with great care, both because of the complex relationship between structures and currents, and the diverse and variable nature of the currents themselves (e.g. Allen, 1966). None the less it has been pointed out that for Recent sedimentary environments there is a close correlation between their current systems and the local and regional orientation of their sedimentary structures (Klein, 1967). Similarly Selley (1968) has suggested that there appear to be a number of palaeocurrent patterns which have been recorded repeatedly from rocks of different ages all over the world. These different patterns or models are specific to various environments. Each can be defined both by the palaeocurrent vector at individual sample stations and by their regional relationships (Table 1.4).

Environment	Local current vector	Regional pattern
Alluvial { Braided	Unimodal, low variability	Often fan-shaped
Meandering	Unimodal, high variability	Slope-controlled, often centripetal basin fill
Eolian	Uni- bi- or polymodal	May swing round over hundreds of miles around high pressure systems
Deltaic,	Unimodal	Regionally radiating
Shorelines and Shelves	Bimodal (due to tidal currents), sometimes unipolar or polymodal	Generally consistently oriented onshore offshore, or longshore
Marine turbidite	Unimodal (some exceptions noted, see p. 189)	Fan-shaped or, on a larger scale, trending into or along trough axes

Table 1.4. A classification of some palaeocurrent patterns. Based on data in Selley (1968).

It would seem therefore that, though it must be interpreted with great care, palaeocurrent analysis is an important technique for recognizing ancient sedimentary environments and their palaeo-geographies.

This approach is severely restricted in sub-surface studies, though some successful applications of dipmeter data in palaeocurrent analyses have been reported (Campbell, 1968; McDaniel, 1968).

Fossils

Last, but certainly not least, of the five parameters defining a facies are its fossils. These have always been one of the most important methods of identifying the depositional environment of a sediment. The way in which fossils lived, behaved towards one another, and influenced and were influenced by their environment is termed palaeoecology. Reviews of this large field of research have been given by Hedgpeth and Ladd (1957), Gecker (1957), Ager (1963), and Imbrie and Newell (1964). To use fossils to identify the depositional environment of the host sediment two assumptions must be made:

(*i*) That the fossil lived in the place where it was buried.

(*ii*) That the habitat of the fossil can be deduced either from its morphology or from studying its living descendants (if there are any).

These are two very real problems which must always be kept in mind when using fossils as environmental indicators. It is not always easy to be sure that a creature lived in or on the sediment in which it was buried. Many fossils are preserved in a particular environment not because they lived in it but because they found their way into it by accident and it was so hostile that it killed them. Think of all the drowned cats washed out to sea by the River Thames.

The second problem, that of deducing the habitat of a fossil, is also a very real one which has been discussed extensively in the literature. To consider one simple case, it has been pointed out that bears live today from the arctic to the equator. If only Polar bears lived at the present time old bear bones might be used as indicators of glacial climates (Shepard, 1964, p. 4).

Lest the preceding paragraphs have painted too gloomy a picture of fossils as environmental indicators it must be stated that they are one of the most valuable techniques available.

Of all the different fossils that can be used in environmental analysis perhaps two of the most important types are micro-fossils and trace fossils. Micro-fossils have the great advantage over megafossils that they are recoverable from well cuttings, and that a small volume of rock may contain sufficient specimens to be used in statistical studies. Trace fossils are at present one of the least understood of all categories of fossils. They have the great advantage, however, that they undoubtedly occur *in situ*. Though the beasts which generated trace fossils are often unknown, many studies of Recent and ancient sediments show that various assemblages of trace fossils are specific to environments and have changed little through geological time (Seilacher, 1967).

Used carefully these and other fossil groups indicate many environmental parameters including depth, temperature, salinity, current turbulence, and climate. This necessarily brief résumé of a complex problem is discussed in greater depth in the references quoted at the beginning of this section.

Sequence and cycles

One of the most interesting attributes of many sedimentary facies is that they tend to be composed of vertical sequences of sub-facies arranged in orderly and predictable motifs. This phenomenon of cyclic sedimentation has been described and discussed in an extensive literature recently reviewed by Duff, Hallam, and Walton (1967).

Because of the complex interplay of the processes which control the deposition of a sedimentary sequence it is seldom that a cyclic motif, if present, is conspicuous. Instead of seeing four sub-facies arranged ABCD, ABCD, ABCD, etc., it is more usual to find something like ABCABDCDAB and so on. It is possible to deduce from the latter sequence that the 'ideal' cycle is ABCD, but this is based on a highly subjective mental process. Within recent years the eyeball method of analysing cyclic sequences has been replaced by statistical techniques.

These include simple methods which, though statistically naïve, can easily be used and understood by the field geologist (e.g. Selley, 1969b). Other, more sophisticated techniques necessitate the use of a computer, but suffer from the assumptions which they have to impose on the geological data (e.g. Merriam, 1967).

Once the cyclic pattern has been extracted from a sedimentary sequence one is faced with the problem of its interpretation.

Cycle-generating processes have been classified into two groups defined by Beerbower (1964, p. 32) as follows:

(i) *Autocyclic mechanisms:* 'are generated in the depositional prism and include such items as channel migration, channel diversion, and bar migration.'

(ii) *Allocyclic mechanisms:* 'result from changes external to the sedimentary unit such as uplift, subsidence, climatic variation, or eustatic change.'

One of the most interesting results of studies of Recent sediments is that they have shown how sub-environments migrate laterally over

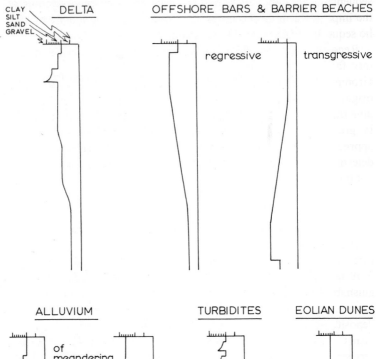

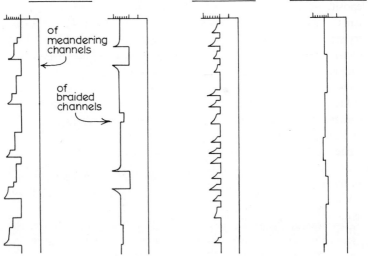

Figure 1.1. Idealized vertical grain size profiles specific to certain environments. No scale. For authentication and explanations, see appropriate chapters in text.

one another across the depositional area resulting in a regular sequence of sub-facies. This has led to an increased understanding of the importance of autocyclic mechanisms in generating ancient cyclic sequences. Previously stress was laid on allocyclic processes.

Recognition of sequences of autocyclic origin within a sedimentary facies is an important way of identifying its depositional environment. Figure 1.1 idealizes the various sequences of autocyclic origin which may be present in different facies. It is interesting to note that these can be detected purely on a basis of vertical variations in grain size without recourse to other facies parameters. This approach can be particularly fruitful in sub-surface studies. The determination of lithology and grain size from electric well logs is not impossible.

SUMMARY

This introductory chapter has covered a great deal of ground so it is perhaps appropriate to close with a summary.

A sedimentary environment is a part of the earth's surface distinguishable from adjacent areas by physical, chemical, and biological parameters. These areas may be erosional, non-depositional, or depositional. Sedimentary facies originate in depositional environments. They may be defined by their geometry, lithology, sedimentary structures, palaeocurrents, and fossils. These reflect not only the depositional environment but also other earlier and time equivalent environments.

The same kind of sedimentary environments occur repeatedly over the face of the earth today. But no two similar environments are ever identical and the boundaries between different adjacent environments are often transitional.

There appear to be a number of ancient sedimentary facies which occur repeatedly in rocks of different ages all over the earth. These can be related to Recent sedimentary environments.

A basic approach to the analysis of ancient sedimentary environments is summarized in Fig. 1.2.

It is generally best to gather all the available data of a sedimentary facies, and not to base the diagnosis of its origin on any single criterion. Geology is at best an imprecise science in which it is seldom possible to make deterministic statements. We deal in probabilities rather than certainties.

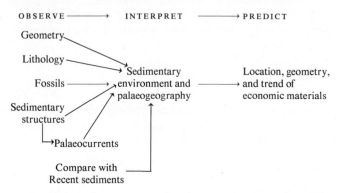

Figure 1.2. Illustrative of the basic approach to finding out how a sediment was deposited.

Finally the student reader must learn to beware of the kind of dogmatic statements with which this chapter abounds. Studious readers may care to immure themselves in the nearest geological library for the next year checking references. Out and out sceptics may prefer to spend the next ten years in the field. Let the rest read on remembering that all generalizations are dangerous including this one.

REFERENCES

Ager, D. V., 1963. *Principles of Paleoecology*. McGraw-Hill, New York. 371 p.

Allen, J. R. L., 1964. The classification of cross-stratified units with notes on their origin. *Sedimentology*, **2**, pp. 93–114.

——, 1966. On bed forms and palaeocurrents. *Sedimentology*, **6**, pp. 153–90.

——, 1967. Depth indicators of clastic sequences. In: Depth indicators in marine sedimentary environments. (Ed. A. Hallam) *Marine Geology*, Sp. Issue, **5**, No. 5/6, pp. 429–46.

——, 1968. *Current Ripples: their relation to patterns of water and sediment motion*. North Holland Pub. Co., Amsterdam, 433 p.

Bathurst, R. G. C., 1967. Depth indicators in sedimentary carbonates. In: Depth indicators in marine sedimentary environments. (Ed. A. Hallam) *Marine Geology, Sp. Issue*, **5**, No. 5/6, pp. 447–72.

ₔerbower, J. R., 1964. Cyclothems and cyclic depositional mechanisms in alluvial plain sedimentation. In: Symposium on Cyclic Sedimentation. (Ed. D. F. Merriam) *Kansas Geol. Surv. Bull.* 169, **1**, pp. 31–42.

Bromley, R. G., 1967. Marine phosphorites as depth indicators. In: Depth indicators in marine sedimentary environment. (Ed. A. Hallam) *Marine Geology*, Sp. Issue, **5**, No. 5/6, pp. 503–10.

Campbell, R. L., 1968. Stratigraphic Applications of Dipmeter Data in Mid continent. *Bull. Amer. Assoc. Petrol. Geol.*, **52**, pp. 1700–19.

Chayes, F., 1956. *Petrographic Modal Analysis. An elementary statistical appraisal.* Wiley & Sons, New York. 113 p.

Conybeare, C. E. B., and Crook, K. A. W., 1968. Manual of sedimentary structures. *Bureau of Mineral Resources.* Bull. No. 102, Canberra A.C.T., 327 p.

Cummings, W. A., 1962. The greywacke problem. *Lpool. Manchr. Geol. J.*, **3**, pp. 51–72.

Degens, E. T., 1965. *Geochemistry of Sediments: a brief survey.* Prentice-Hall, N.J. 342 p.

——, Williams, E. G., and Keith, M. L., 1957. Environmental studies of Carboniferous sediments. In: Geochemical criteria for differentiating marine and freshwater shales. *Bull. Amer. Assoc. Petrol. Geol.*, **42**, pp. 981–97.

Doeglas, D. J., 1946. Interpretation of the results of mechanical analysis. *J. Sediment. Petrol.*, **16**, pp. 19–40.

Duff, P. McL. D., Hallam, A., and Walton, E. K., 1967. *Cyclic Sedimentation.* Elsevier, Amsterdam. 280 p.

Dunham, R. J., 1962. Classification of carbonate rocks according to their depositional texture. In: Classification of carbonate rocks. (Ed. W. E. Ham) *Amer. Assoc. Petrol. Geol.* Memoir No. 1, pp. 108–21.

Folk, R. L., 1967. A review of grain size parameters. *Sedimentology*, **6**, pp. 73–94.

Gecker, R. F., 1957. Introduction to Palaeoecology. French translation, *Bases de la Paleoecologie.* Bur. Récherches Geol. Min. 83 p., also, spelt Hecker, 1965. American Elsevier Pub. Co. N. Yk. 166 p.

Ginsburg, R. N., Lloyd, R. M., Stockman, K. W., and McCallum, J. S., 1963. Shallow water carbonate sediments. In: *The Sea*, vol. III. (Ed. M. N. Hill) Wiley & Sons, N.Y. pp. 554–82.

Grim, R. E., 1958. Concept of diagenesis in argillaceous sediments. *Bull. Amer. Assoc. Petrol. Geol.*, **42**, pp. 246–53.

Gubler, Y. (Ed.) 1966. *Essai de nomenclature et caractérisation des principales structures sédimentaires.* Ed. Tecnip. Paris. 291 p.

Hedgpeth, J. W., and Ladd, H. S. (Eds.) 1957. Treatise on marine ecology and paleoecology, *Mem. Geol. Soc. Amer.* No. 67 (2 vols.) 1296 and 1077 p.

Imbrie, J., and Newell, N. D. (Eds.) 1964. *Approaches to Paleoecology.* Wiley & Sons, N.Y. 432 p.

Imbrie, J., and Purdy, E. G., 1962. Classification of modern carbonate sediments. In: Classification of carbonate rocks. (Ed.
W. E. Ham) *Amer. Assoc. Petrol. Geol.* Memoir No. 1, pp. 253–
73.

Jopling, A. V., and Walker, R. G., 1967. Morphology and origin of ripple-
drift cross-lamination, with examples from the Pleistocene of
Massachusetts. *J. Sediment. Petrol.*, **38**, pp. 971–84.

Krinsley, D. H., and Funnel, B. M., 1965. Environmental history of sand
grains from the Lower and Middle Pleistocene of Norfolk, England.
Quart. J. Geol. Soc. London., **121**, pp. 435–62.

Klein, G. de V., 1967. Paleocurrent analysis in relation to modern marine
sediment dispersal patterns. *Bull. Amer. Assoc. Petrol. Geol.*, **51**, pp.
366–82.

Krumbein, W. C., and Sloss, L. L., 1959. *Principles of Stratigraphy and
Sedimentation.* Freeman, San. Fran. 497 p.

Kuenen, P. H., 1960. Experimental abrasion of sand grains. *Rept. Int. Geol.
Cong.* 21 Session, Pt. 10. Submarine Geology.

McDaniel, G. A., 1968. Application of sedimentary directional features and
scalar properties to Hydrocarbon Exploration. *Bull. Amer. Assoc.
Petrol. Geol.*, **52**, pp. 1689–99.

Merriam, D. F. (Ed.) 1967. *Computer Applications in the Earth Sciences:
Colloquium on time-series analysis. Computer contribution No. 18.
State Geological Survey.* University of Kansas.

Pettijohn, F. J., 1957. *Sedimentary Rocks.* (2nd Edn.) Harper Bros., N.Y.
718 p.

Pettijohn, F. J., and Potter, P. E., 1964. *Atlas and Glossary of Primary
Sedimentary Structures.* Springer-Verlag, N.Y. 370 p.

——, ——, and Siever, R., 1965. *Geology of Sand and Sandstone.* A
conference sponsored by the Illinois Geological Survey and the
Department of Geology, Indiana University. 207 p.

Porrenga, D. H., 1967. Glauconite and chamosite as depth indicators in the
marine environment. In: Depth indicators in marine sedimentary
rocks. (Ed. A. Hallam) *Marine Geology*, Sp. Issue **5**, No. 5/6, pp.
495–502.

Potter, P. E., 1967. Sand bodies and sedimentary environments: a review.
Bull. Amer. Assoc. Petrol. Geol., **51**, pp. 337–65.

——, and Pettijohn, F. J., 1963. *Paleocurrents and Basin Analysis.*
Springer-Verlag, Berlin. 296 p.

——, Shimp, N. F., and Witters, J., 1963. Trace elements in marine and
freshwater argillaceous sediments. *Geochem. Cosmochim. Acta*, **27**,
pp. 669–94.

Rohrlich, V., Price, N. B., and Calvert, S. J., 1969. Chamosite in the Recent
Sediments of Loch Etive, Scotland. *J. Sediment. Petrol.* **39**, pp. 624–
631.

Seilacher, A., 1967. Bathymetry of trace fossils. In: Depth indicators in marine sedimentary environments. (Ed. A. Hallam) *Marine Geology*, Sp. Issue, **5**, No. 5/6, pp. 413–28.

Selley, R. C., 1968. A classification of paleocurrent models. *J. Geol.*, **76**, pp. 99–110.

——, 1969a. Torridonian alluvium and quicksands. *Scott. J. Geol.*, **5**, pp. 328–46.

——, 1969b. Studies of sequence in sediments using a simple mathematical device. *Quart. J. Geol. Soc. Lond.* **125**, pp. 557–81.

Shelton, J. W., 1967. Stratigraphic models and general criteria for recognition of alluvial, barrier-bar, and turbidity current sand deposits. *Bull. Amer. Assoc. Petrol. Geol.* **51**, pp. 2441–60.

Shepard, P. E., 1964. Criteria in modern sediments useful in recognizing ancient sedimentary environments. In: *Deltaic and Shallow Marine Sediments*. (Ed. L. M. J. U. Van Straaten) Elsevier, Amsterdam. pp. 1–25.

Shimp, N. F., Witters, J., Potter, P. E., and Schleicher, J. A., 1969. Distinguishing marine and freshwater muds. *J. geol.*, **77**, pp. 566–80.

Simons, D. B., Richardson, E. V., and Nordin, C. F., 1965. Sedimentary structures generated by flow in alluvial channels. In: Primary sedimentary structures and their hydrodynamic significance. (Ed. G. V. Middleton) *Soc. Econ. Min. Pal.*, Sp. Pub., No. 12, pp. 34–42.

Tanner, W. F., 1967. Ripple mark indices and their uses. *Sedimentology*, **9**, pp. 89–104.

Visher, G. S., 1965. Use of vertical profile in environmental reconstruction. *Bull. Amer. Assoc. Petrol. Geol.*, **49**, pp. 41–61.

Walker, R. G., 1963. Distinctive types of ripple-drift cross-lamination. *Sedimentology*, **2**, pp. 173–88.

Weaver, C. E., 1958. Geologic interpretation of argillaceous sediments. Part I. Origin and significance of clay minerals in sedimentary rocks. *Bull. Amer. Assoc. Petrol. Geol.*, **42**, pp. 254–71.

CHAPTER TWO

RIVER DEPOSITS

INTRODUCTION: RECENT ALLUVIUM

Thick sequences of red-coloured interbedded conglomerates, sands, and shales devoid of marine fossils are common all over the world. These are generally believed to be alluvial in origin. The geomorphology, hydrology, and sedimentology of Recent rivers are well known from both observational and experimental studies (see Leopold, Wolman, and Miller, 1964, and Allen, 1965, for reviews).

Recent alluvial deposits are basically of two kinds whose characteristics are largely related to the morphology of the river channels which deposited them:

(*i*) High-sinuosity meandering channels.

(*ii*) Low-sinuosity braided channel complexes.

Alluvium of meandering rivers

High-sinuosity meandering river channels typically develop where gradients and discharge are relatively low compared to those of braided channel systems (Leopold, Wolman, and Miller, 1964, Fig. 7.39B). At the present time they are characteristic of humid vegetated parts of the world where seasonal discharge rates are fairly steady and sediment availability is relatively low due to subdued topographic relief and the impeding effect of vegetation both on soil erosion and lateral erosion of channel margins. The Mississippi is a good example of a Recent meandering river (Turnbull, Krinitsky, and Johnson, 1950; Fisk, 1944).

The deposits of a meandering river can be sub-divided into three main sub-facies due to deposition in three different sub-environments (Shantser, 1951).

Floodplain sub-facies

Sheets of 'very fine sand, silt and clay are deposited on the overbank areas of the river's floodplain. These are laminated, ripple-marked, and often contain horizons of sand-filled shrinkage cracks, suggesting

sub-aerial exposure. The presence of soils may be indicated by carbonate caliches, ferruginous laterites, and rootlet horizons. Peat may
form and detrital plant debris be preserved on bedding surfaces. This
sub-facies is deposited largely out of suspension during floods when
the river breaks its banks.

Channel sub-facies

The lateral migration of a meandering channel erodes the outer con-
cave bank, scours the river bed, and deposits sediment on the inner

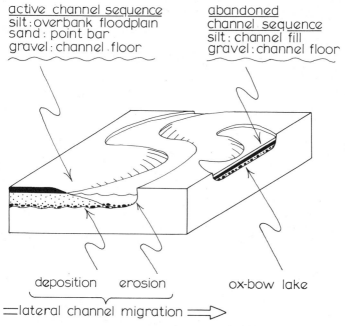

Figure 2.1. Origin of alluvial sub-facies deposited by meandering rivers.

bank (point bar). This produces a characteristic sequence of grain
size and sedimentary structures. At the base is an erosion surface
overlain by extraformational pebbles, intraformational mud pellets,
fragmented bones, and waterlogged drift wood. These originated as a
lag deposit on the channel floor and are overlain by a sequence of
sands with a general vertical decrease in grain size. Massive, flat-
bedded and trough cross-bedded sands grade up into tabular planar
cross-bedded sands of diminishing set height. These in turn pass up

into micro-crosslaminated and flat-bedded fine sands which grade into silts of the floodplain sub-facies (Mackin, 1937; Allen, 1964 Visher, 1965).

Abandoned channel sub-facies
Curved shoestrings of fine-grained deposits infill abandoned channels, sometimes called ox-bow lakes, formed when the river meanders back on itself shortcircuiting the flow. This sub-facies is similar to floodplain deposits but is distinguishable from them by its geometry and because it abruptly overlies a channel lag conglomerate with no intervening point bar sand sequence.

The origin and geometry of the sub-facies deposited by meandering rivers are summarized in Fig. 2.1.

Alluvium of braided rivers

Braided river systems consist of an interlaced network of low sinuosity channels. Recent examples occur on steeper gradients and have higher discharges than meandering rivers (Leopold, Wolman, and Miller, 1964, Fig. 7.39B). Many present-day braided rivers are found on piedmont fans at the edges of mountains where there are large amounts of sediment and discharge is often, but not always, seasonal. Examples have been described from the hot deserts of the mid west of U.S.A. and from the periglacial mountains of the Yukon in Canada (e.g. Blissenbach, 1954, and Williams and Rust, 1969). In these regions erosion is rapid, discharge is sporadic and high, and there is little vegetation to hinder runoff. Because of these factors rivers are generally overloaded with sediment. A channel is no sooner cut than it chokes in its own detritus. This is dumped as bars in the centre of the channel around which two new channels are diverted. Repeated bar formation and channel branching generates a network of braided channels over the whole depositional area. Thus the alluvium of braided rivers is typically composed of sand and gravel channel deposits to the exclusion of fine-grained overbank silts and clays. Due to repeated channel switching and fluctuating discharge there is generally an absence of laterally extensive cyclic sequences similar to those produced by meandering channels. Fining-upward gravel, sand, silt sequences have been recorded however and are attributed to waning current velocities as a channel is gradually infilled (William and Rust, 1969).

The small amount of silt which is present in braided alluvium is generally deposited in abandoned channels. These form both by channel choking and switching (Doeglas, 1962) and by river piracy due to rapid headward erosion of channels on the downslope part of a fan (Denny 1967, Fig. 1). Where an abandoned channel still connects to an active channel downstream it can form a trap for silt and clay carried into it by reverse eddies from the main channel. Thus fine-grained sediments form with shoestring geometries which correspond to the original channel form.

The origin and geometry of the sub-facies deposited by braided rivers are shown in Fig. 2.2.

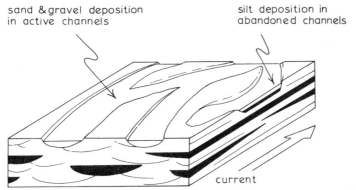

Figure 2.2. Origin of alluvial sub-facies deposited by braided rivers.

Two case histories of ancient alluvium will now be described. In the first of these emphasis will be placed on the deposits of braided rivers, in the second on those of meandering channels.

THE TORRIDON GROUP (PRECAMBRIAN) OF NORTHWEST SCOTLAND: DESCRIPTION AND DISCUSSION

The Torridonian rocks crop out along the northwest coast of Scotland and on adjacent islands of the Hebrides (Fig. 2.3). These are fascinating rocks which have been attributed to nearly every depositional environment and climate known on earth (see Van Houten, 1961). An extraterrestrial origin was proposed by Parkes (1963) who suggested that the Torridonian dropped off the moon. First the facts: these sediments overlie the Lewisian gneiss with an irregular unconformity and are themselves separated by a planar unconformity from overlying Lower Cambrian sandstones and limestones.

The Torridonian is divisible into two groups by an unconformity which separates westerly dipping red sandstones and shales of the Stoer Group from the overlying sub-horizontal Torridon Group. This consists essentially of about 3 km of interbedded red conglomerates, sandstones, and shales. In the north these deposits are separated from the Lewisian gneiss by a planar pre-Torridonian weathered surface. Traced southward, they overlie the Stoer Group. Still further south around the type area of Loch Torridon these deposits overlie and infill a dissected topography cut into the gneiss basement with a relief of several hundred feet. In low-lying parts of this surface breccias banked against the flanks of pre-Torridonian mountain pass laterally into grey shales and sandstones.

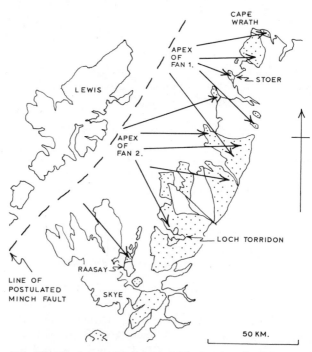

Figure 2.3. Distribution of Torridonian (PreCambrian) sediments, north-west Scotland. Arrows indicate red facies palaeocurrents. Note how these describe two radiating fans whose apices lie along the Minch fault. Modified from Williams, 1969, Fig. 12; by courtesy of the *Journal of Geology*.

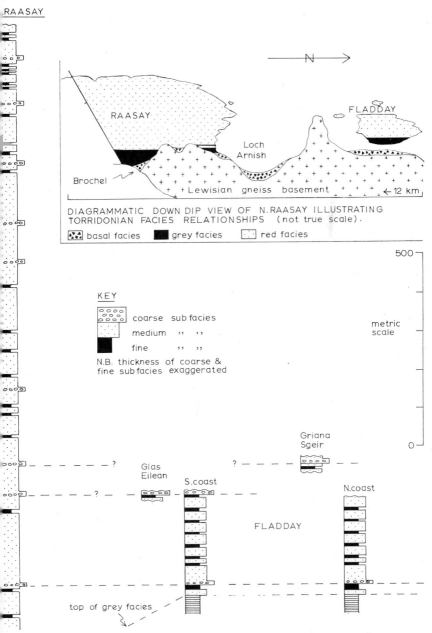

Figure 2.4. Measured sections and facies distribution of Torridon Group sediments on Raasay. Reproduced from Selley, 1969, Fig. 1; by courtesy of the *Scottish Journal of Geology*.

Such a situation can be seen on the island of Raasay whose Torridon Group sediments will now be described. These are divisible into three major facies defined as follows:

(*i*) *Basal facies:* Red and grey breccias and granulestones present adjacent to the buried gneiss topography.

(*ii*) *Grey facies:* Grey sandstones and shales, laterally equivalent to the basal facies.

(*iii*) *Red facies:* Coarse red pebbly sandstones and siltstones overlie the previous two facies.

The regional geometries of the three facies are summarized in Fig. 2.4. Each facies will now be described and its environment discussed.

Basal facies: description

On Raasay the sub-Torridonian unconformity (after allowing for later tectonic tilting) is a dissected plateau with a steep westerly facing scarp some 60 m high. This topography has largely been stripped of its Torridonian cover.

Figure 2.5. Lewisian gneiss boulder in basal facies conglomerate, west coast of Raasay opposite Fladday island. From Selley, 1965, Fig. 6; *Journal of Sedimentary Petrology*; by courtesy of the Society of Economic Paleontologists and Mineralogists.

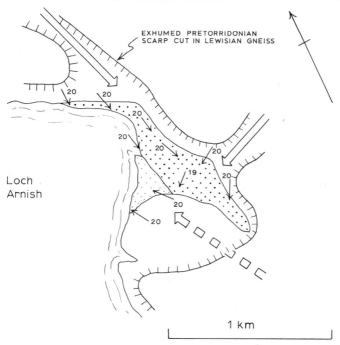

Figure 2.6. Distribution of red basal facies of the Torran outlier, north Raasay. Conglomerates (large dots) overlain by pebbly granulestones (small dots). Small arrows and numbers indicate orientation and inclination of depositional dip, after allowing for tectonic tilting. Thick arrows indicate presumed direction of scree movement down preTorridonian valleys cut in the Lewisian Gneiss. Conspicuous parent valley for upper granulestones absent, but finer grain size suggests a more distant source. From Selley, 1965, Fig. 5, *Journal of Sedimentary Petrology*; by courtesy of the Society of Economic Paleontologists and Mineralogists.

Basal facies deposits can still be found, however, infilling gullies on the scarp crest and as fan-shaped deposits banked against its lower slope.

These sediments are boulder beds, breccias, and granulestones. Individual boulders are up to 2 m in diameter (Fig. 2.5). The composition of the boulders and the arkosic nature of the granulestones leaves no doubt that they are locally derived from the Lewisian gneiss. These sediments are extremely poorly sorted, generally massive, or rudely bedded.

When allowance is made for regional tectonic tilting (by correction using a stereographic net) it can be seen that the bedding of this facies has a considerable depositional dip. In the basal facies on the east side of Loch Arnish dips radiate from exhumed valleys cut in the Lewisian gneiss (Fig. 2.6). These, the stratigraphically lowest basal facies deposits, are red in colour due to a ferric oxide matrix. Higher up the succession, where they pass laterally into the grey facies, the basal facies themselves have a grey-green chloritic cement. Higher still where the basal facies infill gullies on the crest of the Lewisian plateau they are red, like the red facies with which they are laterally equivalent.

Basal facies: interpretation

The coarse grain size, angularity, poor sorting, and petrography of the basal facies deposits clearly show that they were derived locally from the Lewisian gneiss. Their geometry and the radiating depositional dips leaves little doubt that they are ancient piedmont fans. Deposition was probably due to avalanches, mudflows, and sheet floods such as occur on steep slopes near the apices of Recent fans (e.g. Blackwelder, 1928).

Grey facies: description

The basal facies pass laterally into the grey facies at Brochel and along the Raasay coast opposite the island of Fladday. The grey facies consists of over 130 m of three interbedded rock types: thick beds of coarse sandstone, thin beds of medium and fine sandstone, and shales.

Shales are the most abundant rock type, composing about 80 per cent of the grey facies section at Fladday and about 50 per cent at Brochel. These are poorly sorted fissile siltstones, grey in colour, laminated, argillaceous, and micaceous. Thin clay and sand laminae are frequent. Within these shales are lenses and isolated ripples of very fine micro-crosslaminated sands. Desiccation cracks infilled by red medium well-rounded (? wind-blown) sands are common.

Interbedded with the shales are 10–15-cm thick medium and fine-grained sandstones. These are of two types. In the Fladday section are cleanwashed arkoses, micro-crosslaminated from top to bottom. At Brochel, however, equivalent sands are largely poorly sorted vertically graded greywackes. Their bases are often conspicuously erosional and overlain by a thin layer of granules and shale clasts. Internally these sands are generally massive or laminated with occasional convolutions. In contrast to the equivalent beds on Fladday micro-crosslamination is restricted to the top two or three centimetres of these units (Fig. 2.7).

Figure 2.7. Interbedded graded greywackes (?turbidites) and shales with desiccation cracks, overlain by coarse cross-bedded channel sand. Grey facies; Brochel foreshore, east Raasay. From Selley, 1965, Fig. 12, *Journal of Sedimentary Petrology*; by courtesy of the Society of Economic Paleontologists and Mineralogists.

Interbedded with the silts and greywackes at Brochel are coarse grey cross-bedded arkose channel sandstones (Fig. 2.7). Apart from their drab colour these are similar to those of the red facies above.

Grey facies: interpretation

The abundance of laminated grey shale suggests that this facies originated in a low energy environment where fine-grained sediment settled out of suspension. The presence of desiccation cracks shows that the water was shallow and occasionally receded or evaporated.

Intermittent high-energy conditions are indicated by the beds of medium and fine sandstone. The well-sorted, micro-crosslaminated sands of Fladday were probably deposited by gentle traction currents. The poorly-sorted graded greywackes at Brochel, however, show many of the typical features of turbidites (see p. 185).

The coarse cross-bedded channel sands at Brochel herald the advance of the overlying red facies alluvium and, as they debouched into the waters of the grey facies, perhaps generated the turbidites.

Considered overall therefore the grey facies was deposited mostly under quiet shallow water which from time to time receded or evaporated.

Sands, deposited both by traction currents and turbidity flows, indicate higher-energy conditions which ultimately dominated the area due to burial by the red facies. One of the problems of the grey facies, however, is whether it was deposited within enclosed lake basins or in the bays of a dissected coastline. For Phanerozoic rocks this problem would be quickly solved by palaeontology (see p. 70). Though the grey facies shales do contain microfossils their environmental significance is uncertain. Likewise the sedimentological evidence is equivocal. Desiccation cracks can form on both tidal flats and dried out lakes. Turbidity flows have been recorded both from Recent lakes and fjords (p. 187). Palaeocurrents in the grey facies are unipolar, but even bipolar palaeocurrents would not be significant since they have been recorded from ancient lakes and are not therefore exclusive to marine tidal deposits (p. 69). Outcrop is not good enough to show whether the grey facies are restricted to isolated hollows in the gneiss basement or whether they are mutually connected along a dissected shoreline.

Accordingly, for the time being, the marine or non-marine origin of the grey facies of the Torridon Group is unknown.

Red facies: description

The grey and basal facies of the Torridon Group are quantitatively insignificant. The bulk of the sediments are of the red facies which is some 2 km thick on Raasay and over 3 km on the Scottish mainland. The red facies is composed largely of coarse red pebbly arkoses and rare red shales. There are considerable lateral and vertical variations in grain size both on Raasay and adjacent islands and on the mainland.

The red facies on Raasay can be divided into coarse-, medium-, and fine-grained sub-facies defined by their grain size and sedimentary structures.

The coarse sub-facies is composed of very coarse red pebbly arkoses. It occurs in sheets, seldom more than a few metres thick, which can be traced laterally across the southeastwards palaeoslope for kilometres (Fig. 2.4). The bases of these sheets are undulose erosion surfaces overlain by thin intra- and extraformational conglomerates. These pass up into very coarse pebbly sandstones which show flat-bedding and both trough and tabular planar cross-bedding.

Figure 2.8. Ruptured quicksand in red facies sandstone. Foreshore, northwest coast of Fladday island. Reproduced from Selley and others, 1963, plate XVI, Fig. 1; by courtesy of the *Geological Magazine*.

Penecontempraneous deformation can often be seen, including recumbent foresets, convolute bedding, and huge diapiric structures several metres high (Fig. 2.8).

The medium-grained sub-facies makes up the bulk of the red facies section on Raasay, being composed of coarse-, medium-, and fine-grained arkoses. These generally occur in lenticular beds about a metre thick; bounded above and below by red siltstone layers one or two centimetres thick. These sandstones are better sorted than those of the previous sub-facies and pebbles are generally absent. Sedimentary structures include tabular planar cross-bedding and horizontal bedding. These are often deformed in the various manners described for the previous sub-facies. Layers of heavy minerals, mainly iron

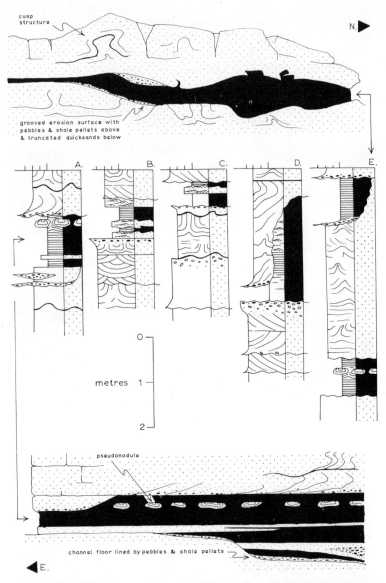

Figure 2.9. Measured sections and sketches of silt-infilled abandoned channels in red facies sandstones, Raasay. From Selley, 1969, Fig. 4; by courtesy of the *Scottish Journal of Geology*.

ore, are common in this and the previous facies. They too are often deformed.

The fine sub-facies makes up about 10 per cent of the total section of the red facies. It is composed of dark red soft weathering shaley argillaceous siltstone with thin interbeds of very fine pink sandstones. The sandstones are often deformed due to penecontemporaneous collapse into the underlying shale forming loadcasts and pseudo-nodules. Internally the sands are generally micro-crosslaminated throughout, with ripples preserved on the upper surface where they are overlain by siltstone. The siltstones are generally laminated and sometimes contain laminae and isolated ripples of sand.

The fine-grained sub-facies occurs in sequences about a metre thick. At the base they overlie laterally extensive scoured surfaces which truncate primary bedding and quicksand structures of the sand beneath. The contact is marked by a thin conglomerate of extra-formational pebbles, mainly of quartz and igneous rocks. The tops of siltstone sequences are also erosional and channelled. The contact with overlying sandstones is marked by a thin conglomerate of both extraformational and intraformational red shale pebbles (Fig. 2.9).

Palaeocurrents determined from cross-bedding in the red facies on Raasay indicate deposition by southeasterly directed currents. There is a very low scatter of readings. Regionally, however, red facies palaeocurrents radiate from two point sources west of the present outcrop (Fig. 2.3). Deposits of the northern fan wedge out south-wards beneath the second.

Red facies: interpretation

The coarse grain size and cross-bedding of this facies indicate deposition by unidirectional high velocity traction currents. The lenticular geometry of the sand beds suggests sedimentation by alternate scouring and infilling of hollows. Shale laminae between sand lenses indicate that current velocity fluctuated widely. The shale lenses between erosion surfaces are clearly abandoned channel deposits. Considered overall therefore the evidence suggests that the red facies originated in an alluvial environment; the sandstones having been deposited from migrating megaripples by violent currents in active channels; and the shales originating in abandoned channels. Nearly half the red facies sandstones were penecontemporaneously deformed. Foresets are overturned downcurrent, original flat-bedding convoluted, whole sandstone beds form diapiric intrusions metres high, and heavy mineral bands founder down into arkose sand. The dominantly vertical trend of these structures suggests that they are not slumps, due to lateral movement, but fossilized quicksands due to water escaping from waterlogged sediment. The expulsion of pore water could have been initiated by seismic vibrations, by turbulent currents, or hydrostatic pressure generated by connate water migrating down the palaeoslope. All these processes have successfully generated analogous quicksand structures in the laboratory. The weight of the evidence suggests however that turbulent currents were the most probable initiator of quicksand movement. Deformational structures similar to those of the red facies (though generally smaller and less abundant) are ubiquitous in sands deposited by traction currents in many environments.

Considering the red facies as a whole, coarse grain size, low palaeocurrent variability, and absence of shale sheets with transitional bases comparable to overbank sediment all suggest deposition by low sinuosity braided channel complexes, rather than by high sinuosity meandering rivers. This explanation is consistent with the regionally radiating palaeocurrents which strongly suggest that the red facies was deposited on piedmont fans. It is interesting to note that the fan apices lie along the line of the Minch fault (evidence for the existence of which is based on a deep, glacially scoured, now subsea, valley and an offset of structural zones between the Lewisian gneiss of the Outer Hebrides and the Scottish mainland (Dearnley, 1962)). This structure has been interpreted as a Caledonian tear fault analogous to that of the Great Glen. It is interesting to speculate, however, that the Minch fault had an easterly downthrow in the late PreCambrian. The Torridon Group piedmont fans may have been generated by its escarpment.

Geopoetry aside, it can be seen that these sediments are a good example of ancient alluvium attributable to deposition from braided rivers.

DEVONIAN SEDIMENTS OF SOUTH WALES AND THE CATSKILL MOUNTAINS, U.S.A.: DESCRIPTION

At the end of the Silurian Period the Caledonian orogeny formed extensive mountain chains in the continental areas around the North Atlantic. Marginal basins, often fault-bounded, were infilled during the Devonian Period by thick sequences of red conglomerates, sand-stones, and shales (the Old Red Sandstone). These thin away from the mountainous source regions and pass into fine-grained marine sedi-ments. Red-bed sedimentation was generally terminated by a marine transgression in the Lower Carboniferous. Devonian red beds have been intensively studied in North America, Greenland, Spitzbergen, and the United Kingdom.

Nomenclature		Sedimentology	Fauna	Environment
N.E. U.S.A.	S. Wales			
Pocono	Facies A	Coarse red pebbly cross-bedded sandstone, rare shales	Generally barren	Braided alluvium
Catskill	Facies B	Interbedded red medium sands and shales	Fish, lamellibranchs plants	Meandering alluvium
Chemung	Facies C	Interbedded grey fine sands and shales, rare coals	Plant debris, burrows, lamellibranchs, and brachiopods	Marine shoreline
Portage	—	Laminated dark pyritic shales, rare dark calcilutites and greywackes	Ammonoids lamellibranchs	Open marine

Table 2.1. Summary of Devonian facies in south Wales and Pennsylvania and New York State, U.S.A.

Based on data from northeastern U.S.A. and south Wales four main Devonian facies can be recognized (Table 2.1). Their stratigraphy is summarized in Fig. 2.10, and their outcrops in Figs. 2.11 and 2.12.

The Catskill facies (facies B of south Wales) will now be de-scribed. The reasons why it is thought to be the alluvium of meander-ing rivers will then be discussed.

In its type area of the Catskill Mountains this facies has a north-westerly thinning prismatic geometry with a maximum thickness of about 2,000 ft. It is overlain by the westerly diachronous Pocono facies and itself diachronously overlies the Chemung facies. The analogous facies 'B' deposits of south Wales show a more erratic and

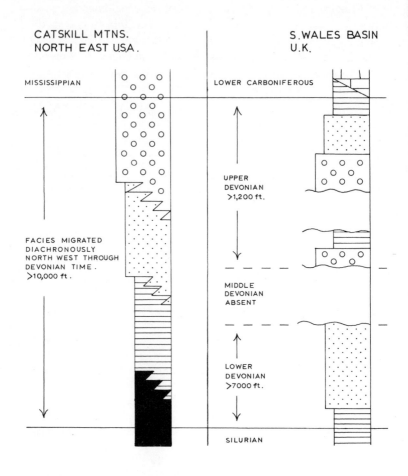

Figure 2.10. Generalized comparative sections of Catskill (U.S.A.), and south Wales Devonian sediments.

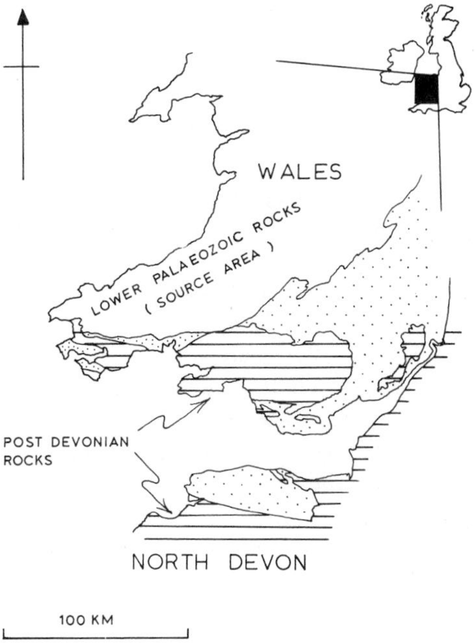

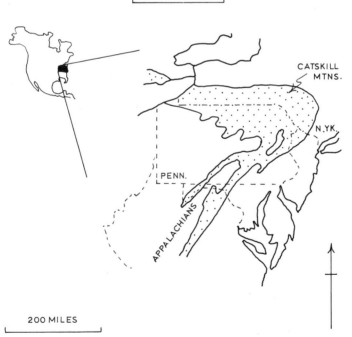

Figure 2.11. Map of south Wales and adjacent areas showing outcrop of Devonian strata (stippled). Almost totally continental (Old Red Sandstone) facies in south Wales pass southwards into interbedded marine and continental facies in north Devon.

WALES

LOWER PALAEOZOIC ROCKS (SOURCE AREA)

POST DEVONIAN ROCKS

NORTH DEVON

100 KM

CATSKILL MTNS.

N.YK.

PENN.

APPALACHIANS

200 MILES

Figure 2.12. Outcrop of Devonian sediments, northeastern U.S.A.

less well-defined distribution. They occur at two stratigraphic levels being limited vertically both by unconformities and changes into Pocono- and Chemung-type sediments (Fig. 2.10).

Lithologically the Catskill facies consists of sandstones and shales in about equal proportions with a minor amount of conglomerates.

The conglomerates are thin, lenticular, and seldom more than a few decimetres thick. They are composed of extraformational pebbles of vein quartz and rock fragments, and of intraformational shale pebbles of local penecontemporaneous origin.

The sandstones are variable in texture and composition. Generally they are medium- to fine-grained and seldom well sorted. A sparse clay matrix is sometimes present. Petrographically these vary from lithic sandstones to sub-greywackes. In south Wales they are red-coloured due to ferric iron in the matrix. In the Catskill region they are generally drab or mottled.

The shales with which the sandstones are interbedded are argillaceous sandy siltstones with thin fine sand layers. A red colour predominates though drab and mottled units do occur. Interbedded with the shales are occasional thin layers of nodular microcrystalline limestone and dolomite. Pebbles of this lithology sometimes occur in the conglomerates due to re-working.

The three lithologies of the Catskill facies have a tendency to be repeatedly arranged in upward-fining sequences of conglomerate, sand, and shale. These lithologies are accompanied by a corresponding systematic vertical arrangement of sedimentary structures (Fig. 2.13).

Each sequence commences with an erosional, scoured, and often channelled surface cut into the shales beneath. This is overlain by the thin conglomerate member, though it may locally be absent. The conglomerate is generally massive or sometimes poorly stratified with bedding surfaces arranged sub-parallel to channel margins.

The sandstones above the conglomerates are generally medium-grained and cross-bedded in their lower part grading up into fine flat-bedded and micro-crosslaminated sandstone. This unit in turn passes up transitionally into the shales. These generally show lamination and contain thin sandstone layers, ripples, sand-filled desiccation cracks, and nodular carbonate horizons. These fining-upward sequences of conglomerate sand and shale are generally between 2–15 m thick. The precise geometry of these sediments is ill-defined due to poor exposure.

Dip directions of cross-bedding are highly variable at individual sample points. Plotted regionally, however, they generally show a preferred trend which is believed to correspond to the palaeoslope. This was southwards in the Welsh basin and generally northwest-wards in the Catskills.

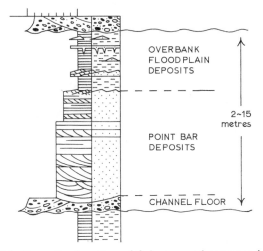

Figure 2.13. Generalized section of fining-upward sequence in Devonian Catskill facies, U.S.A. and south Wales.

Fossils in this facies are generally hard to find. Shales sometimes contain plant fragments and rootlet horizons. The sandstones occasionally yield disarticulated plates and spines of primitive fish such as the Pteraspids and lamellibranch shells often, in south Wales, called *Archanodon jukesi*. Plant fragments are sometimes present in the sandstones and conglomerates.

DEVONIAN SEDIMENTS OF SOUTH WALES AND THE CATSKILL MOUNTAINS, U.S.A.: INTERPRETATION

The repeated upward-fining of grain size in this facies suggests deposition by currents which at any one point waxed quickly and waned slowly in velocity through time. Thus the conglomerate-covered erosion surfaces record scouring turbulent currents while the overlying sands indicate lower velocity unidirectional traction currents which deposited cross-beds from migrating megaripples.

The gradation up into siltstone shows that velocity diminished to the extent that fine sediment could settle out of suspension. Associated desiccation cracks, plant roots, and nodular limestones, interpreted as caliches, indicate intermittent sub-aerial exposure.

The fauna is generally believed to be freshwater. The reason for this is largely a circular one: that the deposits are continental, therefore the fossils are non-marine. It is significant, however, that there is an absence of undoubtedly marine fossils, such as brachiopods, echinoderms, and trilobites. Similarly the occurrence of plant debris and rootlet horizons suggests that the environment, if not continental, was certainly near shore and intermittently swampy.

Considered altogether the evidence points to the conclusion that the Catskill deposits were laid down in a continental environment. The vertical sequence of lithologies and sedimentary structures is closely comparable to those produced today by the lateral migration of meandering rivers (p. 23); the conglomeratic scoured surfaces originating on channel floors, the sands being deposited on the point bars and the silts on the floodplains (Mackin, 1937; Allen, 1964; Visher, 1965).

This type of fining-upward cycle occurs in many ancient alluvial deposits (see Allen, 1965, for a review). Before this simple explanation was advanced several other suggestions had been made. These included eustatic changes of sea level, which would affect the base level and equilibrium of a river; tectonic uplift of the source, which would increase sediment supply; and climatic changes varying rainfall and hence runoff and current velocity.

It is clear, however, that the to and fro migration of a river across its floodplain is an adequate explanation for alluvial cycles when superimposed on gradual tectonic subsidence. This is not to say that the other processes did not affect sedimentation but that if they did then their effects must be superimposed upon this built-in cyclic mechanism of an alluvial floodplain. Certainly the large-scale vertical facies variations and unconformities of the south Wales basin indicate forces beyond those generated by a meandering river. Similarly five upward-fining sequences occur in the Pleistocene and Recent alluvium of the Mississippi river valley. These can be attributed to eustatic sea-level changes during the ice age (Turnbull, Krinitsky, and Johnson, 1957). For further discussion of the origin of cycles in alluvial successions see Beerbower (1964), Allen (1964; 1965), Visher, (1965), and Duff, Hallam, and Walton (1967, pp. 21–35).

DISCUSSION

Large volumes of rock in many parts of the world ranging in age from PreCambrian to the present day have been attributed to alluvial environments. Basically these can be classified into four main types:

(*i*) Prisms of sediment thousands of feet thick deposited in basins adjacent to mountain chains.

(*ii*) Sequences, thousands of feet thick, deposited in fault-bounded troughs within continental shields.

(*iii*) Laterally extensive sheets of coarse braided alluvium generally only a few hundred feet thick blanketing continental shields.

(*iv*) Thin sheets of alluvium beneath transgressive marine deposits.

The first type of alluvial deposit is shown by the Catskill and associated facies of North America and by the south Wales Old Red Sandstone basin just described. Other examples include the Upper Cretaceous sediments east of the Rocky Mountains (Chapter 6), the Tertiary molasse of Switzerland, and the Siwalik Series south of the Himalayas. These examples have the following features in common. They all were formed in basins along the edges of rising mountain chains. They are thousands of feet thick adjacent to the mountains from which they were derived. The sediments thin away from the source as grain size diminishes. Facies analysis suggests that the depositional environment changed from braided alluvium at the source, through meandering alluvium to marine shoreline and, ultimately, open marine conditions.

The second type of alluvium occurs in fault-bounded troughs often closed from the sea either within mountain chains or on continental shields. These deposits, often thousands of feet thick, are generally coarse braided alluvium generated from fans on the marginal fault scarps. They may pass laterally into meandering alluvium or directly into lacustrine deposits in the trough centre. These deposits are often interbedded with volcanic rocks which erupted along the marginal faults. Examples of this model include the Triassic Newark: Connecticut trough complex of northeast North America (Klein, 1962; Van Houten, 1964), the Devonian Midland Valley of Scotland, the PreCambrian Dala and Jotnian sandstones of Scandinavia, and, possibly, the Torridonian.

The third type of alluvial occurrence are sheets of coarse pebbly sandstone which, though only a few hundred feet thick, cover hundreds of square miles. At their base they are separated by a conglomerate from a planar unconformity though in detail this may be channelled or with protruding hillocks.

Extensive deposits of this type occur around the edges of the Saharan and Arabian Shield. These range in age from Cambrian to Recent (Picard, 1943), and are loosely referred to as of Nubian facies (the type Nubian is Cretaceous). Sedimentologically these rocks are closely comparable to the braided alluvium of the Torridonian red facies. It is difficult to see, however, how the steep gradients and high current velocity necessary for braided rivers could have been maintained downcurrent over hundreds of miles. One would expect sediment to quickly build up to base level so that current velocity diminished and fine-grained meandering alluvium deposited.

An answer to this problem was suggested by Stokes (1950) in a study of the Shinarump and similar formations on the Colorado plateau. Here there are several sheets of coarse sandstones and conglomerates each generally less than 300 ft thick and overlying a planar unconformity. Stokes points out that, since these rocks contain terrestrial fossils, it is unlikely that they were laid down by a sea transgressing over a peneplain. It is more probable that these sand sheets were deposited on piedmont fans derived from scarps which retreated as they cut back across a pediplain. Williams (1969) put forward a similar explanation for the Torridonian fans.

This concept can be applied to the Nubian facies of the Saharan and Arabian Shields. Typical examples of these deposits occur in the Southern Desert of Jordan (Fig. 2.14). Here the PreCambrian igneous basement is overlain by 700 m of coarse cross-bedded pebbly sandstones attributable to a braided alluvial environment (Selley, 1970; 1972). The alluvial facies is divisible into three formations (Fig. 2.15). The unconformity with the PreCambrian basement is a planar (?penecontemporaneously) weathered surface south of Wadi Rum. Downcurrent to the north inselbergs protrude 35 m into the overlying Saleb Formation which thickens northwards from 30 m south of Wadi Rum to about 60 m in a distance of 30 km. The base of the overlying Ishrin Formation is marked by huge channel complexes whose floors are lined with imbricate siltstone slabs over 1 m long. These were presumably re-worked from the Saleb Formation beneath. The Ishrin Formation is about 300 m thick and shows little regional thinning over hundreds of square miles. The top of the Ishrin Formation is incised by channels over 5 m deep which again are sometimes completely infilled by siltstone slabs. These channels mark the base of the Disi Formation which, like the Ishrin Formation, is about 300 m thick and shows little regional thickness variation over hundreds of square miles. At its top the Disi passes abruptly, but apparently without erosion, into overlying marine shelf sands of the Um Sahm Formation (p. 154).

Figure 2.14. Pediment cut in PreCambrian igneous basement (dark rock), overlain by Cambro-Ordovician braided alluvium of the Saleb and Ishrin formations. Recent braided alluvial fan in foreground. Wadi Rum, Jordan.

Deposition from the braided fans of repeatedly retreating scarps is certainly a most attractive explanation for these extensive alluvial deposits. According to this concept the Arabian Shield was uplifted three times. Each phase of rejuvenation of the landscape caused scarps to retreat into the hinterland of the shield. The first scarp would have cut a pediment into the basement on which the braided alluvium of the Saleb Formation was deposited. The next phase of uplift caused a second scarp to cut a new pediment into the Saleb deposits and bury it with Ishrin alluvium. A third repetition of this process deposited the Disi Formation (Fig. 2.15).

This process can explain the Nubian-type sand sheets of the Arabian and Saharan shields.

The fourth type of alluvial deposit, thin sheets beneath marine transgressions, is genetically related to shorelines and discussed therefore in Chapter 6.

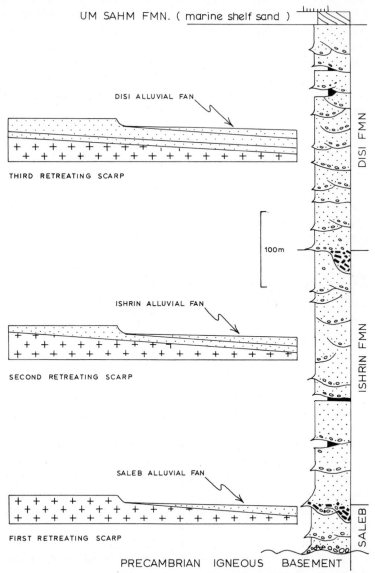

UM SAHM FMN. (marine shelf sand)

DISI ALLUVIAL FAN

THIRD RETREATING SCARP

DISI FMN

100m

ISHRIN ALLUVIAL FAN

SECOND RETREATING SCARP

ISHRIN FMN

SALEB ALLUVIAL FAN

FIRST RETREATING SCARP

SALEB

PRECAMBRIAN IGNEOUS BASEMENT

Figure 2.15. Section of Cambro-Ordovician braided alluvial sandstones in the Southern Desert, Jordan. Diagrams illustrate possible depositional mechanism of repeated uplift and pedimentation.

ECONOMIC ASPECTS

Alluvial deposits are of economic interest in the search for oil, gas, metals, and coal.

In general fluviatile sediments are not very good oil and gas reservoirs. This is because they often form in closed continental basins where there are no source rocks or, if they occur around marine basins, they often lack impervious caprocks. Exceptions to this general rule occur where alluvial sands underlie transgressive marine deposits and where alluvial fans flank rapidly subsiding marine basins. The first situation is discussed with linear clastic shoreline oil reservoirs to which it is genetically related (Chapter 6, p. 110). The second situation is found in the Sirte basin of Libya. Here Lower Cretaceous alluvial fans are banked against an irregular (? faulted) basement topography. These were buried by marine shales of the Cenomanian transgression. Where structurally high relative to the shales, these fan deposits, still largely uncemented, contain some interesting oil fields.

Aside from oil and gas, alluvial deposits can also be metalliferous. The gold deposits of Witwatersrand in South Africa are a case in point. These have been extensively described and discussed in the literature (e.g. Haughton, 1964). The Rand basin lies on the PreCambrian basement of South Africa. It covers some 15,000 square miles and measures about 150 miles from northeast to southwest and 60 miles from northwest to southeast. It is infilled by over 25,000 ft of PreCambrian clastics which coarsen upwards and northwestwards towards their presumed source. There has been considerable discussion about whether the gold formed at the same time as the sediments as a detrital placer deposit or whether it is of later date. There is general agreement however that the ore is concentrated in conglomerate channels which are often up to 2,000 ft wide and 200 ft deep. These curve westwards across the basin from north to south. There appears to be little doubt that the conglomerates originated in a complex of braided river channels.

A second notable example of minerals in alluvium are the Uranium carnotite ores of the Colorado Plateau of the U.S.A. (Nininger, 1956; Lowell, 1955). These occur in channels in Mesozoic red beds. Again there has been considerable discussion of the origin of the ore, but general agreement that its present distribution is sedimentologically controlled. The Mesozoic sediments of the

Colorado plateau consist of a stratigraphically complex series of interbedded alluvial and eolian formations (p. 52). Many of the alluvial deposits originated as piedmont fans. Uranium ore bodies occur in channels at the unconformities at the bases of formations and within channels enclosed in alluvial sequences. Experience has shown that the most favourable location for the ores is often about half-way down the fan where the sand : shale ratio approximates to 50 : 50. Once located the trend of individual ore bodies can be predicted by palaeocurrent analysis of the host channel and associated deposits.

In addition to oil, gas, and other minerals alluvium is associated with coals in deltas and lacustrine swamps. These are discussed on p. 89 and p. 72 respectively.

REFERENCES

The description of the Torridonian sediments was based on:

Dearnley, R., 1962. An outline of the Lewisian complex of the Outer Hebrides in relation to that of the Scottish mainland. *Quart J. Geol. Soc. Lond.*, **118**, pp. 143–66.

Parkes, L. R., 1963. The Origin of the Moon. *New Scientist*, **20**, No. 365, p. 404.

Selley, R. C., 1964. The Penecontemporaneous deformation of heavy mineral bands in the Torridonian Sandstone of N.W. Scotland. In: *Deltaic and Shallow Marine Sediments*. (Ed. L. M. J. U. Van Straaten) Elsevier, Amsterdam. pp. 362–7.

——, 1965a. Diagnostic characters of fluviatile sediments of the Torridonian Formation (PreCambrian) of North-West Scotland. *J. Sediment. Petrol.*, **35**, pp. 366–80.

——, 1965b. The Torridonian Succession on the Islands of Fladday, Raasay and Scalpay, Inverness-shire. *Geol. Mag.*, **102**, pp. 361–9.

——, 1966. Petrography of the Torridonian rocks of Raasay and Scalpay, Inverness-shire. *Proc. Geol. Ass.*, **77**, pp. 293–314.

——, 1969. Torridonian Alluvium and Quicksands. *Scott. J. Geol.*, **5**, No. 4, pp. 328–46.

——, and Shearman, D. J., 1962. Experimental Production of Quicksands. *Proc. Geol. Soc.*, *London*, No. 1599, pp. 101–2.

——, ——, Sutton, J., and Watson, J., 1963. Some Underwater Disturbances in the Torridonian of Skye and Raasay. *Geol. Mag.*, **100**, pp. 224–43.

——, and Stewart, A. D., 1967. The Torridonian Sediments of N.W. Scotland. *Field Guide for Excursion C.2. 7th Internat. Sed. Congress.*

Stewart, A. D., 1966. An Unconformity in the Torridonian. *Geol. Mag.*, **103**, pp. 462–5.

Van Houten, F. B., 1961. Climatic Significance of Red Beds. In: *Descriptive Palaeoclimatology*. (Ed. A. E. M. Nairn) Interscience, N.Y. pp. 89–139.

Williams, G. E., 1969. Characteristics and Origin of a PreCambrian Pediment. *J. Geol.*, **77**, p. 183–207.

The account of the Devonian deposits of the Catskills, U.S.A., and the Old Red Sandstone of south Wales was based on:

Allen, J. R. L., 1964. Studies in fluviatile sedimentation: six cyclothems from the Lower Old Red Sandstone, Anglo-Welsh basin. *Sedimentology*, **3**, pp. 163–98.

——, 1965. Upper Old Red Sandstone Paleogeography in South Wales and the Welsh Borderland. *J. Sediment. Petrol.*, **35**, pp. 167–96.

——, and Friend, P. F., 1968. Deposition of the Catskill Facies, Appalachian Region: with notes on some other Old Red Sandstone Basins. In: Late Palaeozoic and Mesozoic Continental Sedimentation. (Ed. G. de V. Klein) *Geol. Soc. Amer. Sp. Pap.*, No. 206, pp. 21–74.

Friedman, G. M., and Johnson, K. G., 1966. The Devonian Catskill Deltaic Complex of New York, type example of a 'tectonic delta complex'. In: *Deltas*. (Ed. M. L. Shirley and J. A. Ragsdale) Houston Geol. Soc., pp. 171–88.

——, and ——, 1969. The Tully Clastic Correlatives (Upper Devonian) of New York State: A Model for Recognition of Alluvial, Dune (?), Tidal, Nearshore (Bar and Lagoon), and Offshore Sedimentary Environments in a Tectonic Delta Complex. *J. Sediment. Petrol.*, **39**, pp. 451–85.

Friend, P. F., 1966. Clay fractions and colours of some Devonian Red beds in the Catskill Mountains, U.S.A. *Quart. J. Geol. Soc. Lond.*, **122**, pp. 273–92.

Other references quoted in this chapter were:

Allen, J. R. L., 1965a. Fining upward cycles in alluvial succession. *Lpool. Manchr. Geol. J.*, **4**, pp. 229–46.

——, 1965b. A review of the origin and characteristics of Recent alluvial sediments. *Sedimentology*, **5**, No. 2, Sp. Issue. pp. 89–191.

Beerbower, J. R., 1964. Cyclothems and cyclic depositional mechanisms in alluvial plain sedimentation. In: Symposium on Cyclic Sedimentation. (Ed. D. F. Merriam) *Kansas Geol. Surv. Bull.*, 169, **1**, pp. 31–42.

Blackwelder, E., 1928. Mudflow as a Geologic Agent in Semi-arid Mountains. *Bull. Geol. Soc. Amer.*, **39**, pp. 465–83.

Blissenbach, E., 1954. Geology of Alluvial Fans in semi-arid Regions. *Bull. Geol. Soc. Amer.*, **65**, pp. 175–90.

Doeglas, D. J., 1962. The structure of sedimentary deposits of braided rivers. *Sedimentology*, **1**, pp. 167–90.

Denny, C. S., 1967. Fans and Pediments. *Am. J. Sci.*, **265**, pp. 81–105.

Duff, P. McL. D., Hallam, A., and Walton, E. K., 1967. *Cyclic Sedimentation*. Elsevier, Amsterdam. 280 p.

Fisk, H. N., 1944. Geological investigation of the Alluvial Valley of the Lower Mississippi River. *Mississippi River Commission, Vicksburg.* 78 p.

Haughton, S. A. (Ed.), 1964. The geology of some Ore Deposits in Southern Africa. *Geol. Soc., South Africa*, 2 vols.

Klein, G. de V., 1962. Triassic Sedimentation, Maritime Provinces, Canada. *Bull. Geol. Soc. Amer.*, **73**, pp. 1127–46.

Leopold, L. B., Wolman, M. G., and Miller, J. P., 1964. *Fluvial Processes in Geomorphology*. Freeman, San. Fran. 522 p.

Lowell, J. D., 1955. Applications of Cross-stratification studies to problems of Uranium Exploration, Chuska Mountains, Arizona. *Econ. Geol.*, **50**, pp. 177–85.

Mackin, J. H., 1937. Erosional History of the Big Horn Basin, Wyoming. *Bull. Geol. Soc. Amer.*, **48**, pp. 813–94.

Nininger, R. D. (Ed.), 1956. *Exploration for Nuclear Raw Materials*. Macmillan, London. 293 p.

Picard, L., 1943. *Structure and Evolution of Palestine*. Geol. Dept. Hebr. Univ., Jerusalem. 134 p.

Selley, R. C., 1970. Ichnology of Palaeozoic sandstones in the Southern Desert of Jordan: a study of trace fossils in their sedimentologic context. In: *Trace Fossils*. (Harper, J. C., and Crimes, T. P., Eds.) Lpool. geol. Soc. pp. 477–88.

——, 1972. Diagnosis of marine and non-marine environments from the Cambro-ordovician sandstones of Jordan. *Jl. geol. Soc. Lond.* **128**, pp. 109–17.

Shantser, E. V., 1951. Alluvium of river plains in a temperate zone and its significance for understanding the laws governing the structure and formation of alluvial suites. *Akad. Nauk. S.S.S.R. Geol. Ser.* **135**, pp. 1–271.

Stokes, W. L., 1950. Pediment Concept Applied to Shinarump and similar Conglomerates. *Bull. Geol. Soc. Amer.*, **61**, pp. 91–8.

Turnbull, W. J., Krinitsky, E. S., and Johnson, L. J., 1950. Sedimentary Geology of the Alluvial Valley of the Mississippi River and its bearing on foundation problems. In: *Applied Sedimentation*. (Ed. P. D. Trask) Wiley, N.Y. pp. 210–26.

Van Houten, F. B., 1964. Cyclic Lacustrine Sedimentation, Upper Triassic Lockatong Formation. Central New Jersey and adjacent Pennsylvania. In: Symposium on Cyclic Sedimentation. (Ed. D. F. Merriam) *Geol. Surv. Kansas Bull.*, 169, **2**, pp. 497–532.

Visher, G. S., 1965. Fluvial Processes as interpreted from ancient and recent fluviatile deposits. *Soc. Econ. Pal. Min.* Sp. Pub., No. 12, pp. 116–32.

Williams, P. F., and Rust, B. R., 1969. The Sedimentology of a braided river. *J. Sediment. Petrol.*, **39**, pp. 649–79.

CHAPTER THREE

WIND-BLOWN SEDIMENTS

INTRODUCTION

Ancient wind-deposited sediments are often very hard to distinguish from those laid down by running water. This is because the mechanics of both processes are often similar, with transportation and deposition taking place by migrating ripples and dunes.

As this chapter shows, however, a number of criteria have been proposed to distinguish eolian from aqueous sediments. Some of these are of debatable merit when used in isolation. The ultimate decision of whether a sediment is of eolian origin must be based on a critical evaluation of all the available data. Fortunately this is helped by studies of Recent wind-blown deposits, notably by Bagnold (1941) and McKee (1966).

Ancient rocks which have been attributed to eolian deposition are quantitatively insignificant in the sedimentary rocks of the world, but they are widely distributed and range in age from PreCambrian to Recent. Some of the best documented complexes of supposed ancient eolian sediments occur in the western U.S.A. These will now be briefly described and the reasons for attributing them to eolian action stated.

EOLIAN DEPOSITS OF WESTERN U.S.A.: DESCRIPTION

The Colorado Plateau occupied parts of the states of Colorado, Arizona, Utah, and New Mexico in western U.S.A. (Fig. 3.1). This region was an important site of intermittent eolian sedimentation through some 150 m.y. from the Pennsylvanian (Upper Carboniferous) to the Upper Jurassic. The regional geology is complicated but can often be resolved due to a combination of good exposure and dissected topography (Fig. 3.2). In general the Cordilleran geosyncline lay to the west and was an area of marine deposition through much of this time. The present Colorado Plateau was then an unstable shelf on which sedimentation took place in a number of basins which were intermittently connected to one another and to the sea. These include the Paradox, San Juan, Kaiparowits, and Black Mesa basins. The stratigraphy of this region is complex therefore, with rapid lateral and vertical facies changes due to both tectonic and

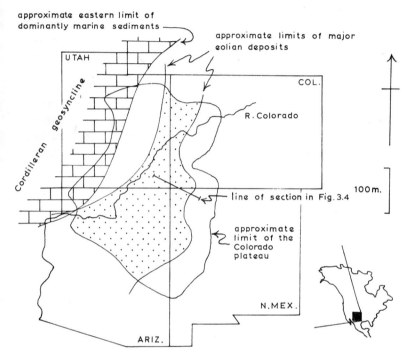

Figure 3.1. Areal distribution of ancient eolian sandstones of the Colorado plateau, U.S.A.

sedimentary causes. Three main facies can be recognized in the Pennsylvanian to Jurassic rocks of the Colorado Plateau:

(*i*) Easterly thinning sheets of limestones, shales, dolomites, and evaporites (thought to be marine).

(*ii*) Wedges and fans of red sandstones, siltstones, and conglomerates, generally best developed in the east, e.g. the Moenkopi and Morrison Formations (thought to be fluviatile).

(*iii*) Red and white sandstones with irregular sheet geometries, e.g. the Entrada, Navajo, Wingate, Weber, De Chelly, and Coconino Formations (thought to be eolian).

Figure 3.2. Isolated butte illustrating good exposure of the Colorado plateau. Monument Valley, Arizona. Massive Wingate eolian sandstone (Trias–Jurassic) forms a steep cliff above a talus slope with ledges of fluviatile Chinle sandstones and shales (Trias). Photo by courtesy of W. F. Tanner.

The supposed eolian sandstones are interbedded with the other two facies with interfingering but generally abrupt margins. They are irregular in plan and seldom greater than 200 ft thick though some examples like the Navajo Sandstone approach 1,000 ft. Lithologically these deposits are exclusively sand-grade sediment. Petrographically they are protoquartzites with minor amounts of feldspar, mica, chert, and red ferruginous clay. Cementation is by quartz grain overgrowths and microcrystalline intergranular quartz and calcite. In one or two exceptional cases, as in the upper part of the Todilto Formation, wind-blown gypsum sands occur. The grade of the sandstones ranges from very fine to coarse, but it is mainly fine. Sorting is moderate to good, occasionally bimodal, with coarse grains in a fine sand. Rounding of grains is moderate to good.

Cross-bedding is the characteristic sedimentary structure of these rocks (Fig. 3.3). Both tabular planar and trough sets are present. Set heights vary from 1–100 ft and troughs range in width from 2–100 ft. Individual foresets dip at between 20–30° and are generally curved at the base. Low angle backsets also occur sub-parallel to erosion surfaces. Penecontemporaneous deformation of bedding is sometimes present. Some of the fine sandstones show long wavelength low amplitude asymmetrical ripples with R.I.s between 20 and 50 (the ripple index (R.I.) is calculated by dividing the wavelength by the amplitude).

Dip directions of cross-bedding show wide scatters within one outcrop. Throughout the majority of Pennsylvanian to Upper Jurassic time they indicate transport generally to the south, with some variation to southeasterly and southwesterly directions.

Figure 3.3. Eolian cross-bedding in Entrada Formation (Jurassic), Church rock, New Mexico. Photo by courtesy of W. F. Tanner.

These sandstones are devoid of fossils except for occasional foot-prints attributed to terrestrial quadrupeds and bipeds, largely dino-saurs.

EOLIAN SANDSTONES OF WESTERN U.S.A.: DISCUSSION

There are two main lines of argument for the eolian origin of these sands. Some arguments are negative, suggesting that these are not water-laid sands, others are positive, suggesting that they are wind-blown.

It is unlikely that these are water-laid sands because of the absence of pebbles, which are generally too heavy to be wind-blown, and of clay, which is generally too light to come to rest on a windy earth. There is no sign of an aqueous biota either marine or non-marine. Conglomerate-lined channels suggestive of running water are absent.

Positively in favour of an eolian origin for these sandstones is their close comparison with Recent desert dunes. Points in common are the dominance of sand-grade sediment which is generally well sorted, matrix free, and well-rounded. Cross-bedding of such vast height is unknown in Recent aqueous sediments, but is known from Recent terrestrial dunes. Ripple indices >15 are recorded from Recent eolian ripples whereas aqueous ripples have R.I.s <15 (Tanner, 1967). Recent dune country is generally lifeless, such plants and animals as live and die there are desiccated and destroyed by the shifting sands. The persistent southerly direction of sand transport is perpendicular to the (general) westerly palaeoslope of the Colorado Plateau and locally can be seen trending oblique to slope-controlled palaeocurrents of the associated fluviatile red beds (Fig. 3.4).

In conclusion, therefore, an eolian origin for these sandstones of the Colorado Plateau is strongly indicated by the absence of features indicating aqueous deposition and by their close similarity to Recent eolian dunes. Lingering scepticism for this explanation can generally be allayed by asking what other sedimentary environment could generate the features seen in these sandstones.

Due to the wealth of stratigraphic and sedimentologic data avail-able it is possible to reconstruct the palaeogeography of the ancient Colorado Plateau dune fields in some detail. In Pennsylvanian and early Permian time dunes extended intermittently along a coastal plain separating the Rocky Mountain geosynclinal seas from alluvial

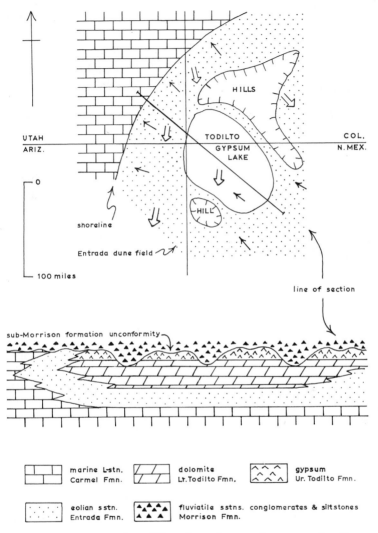

UTAH / ARIZ.

COL. / N.MEX.

HILLS

TODILTO GYPSUM LAKE

HILL

shoreline

Entrada dune field

line of section

0

100 miles

sub-Morrison formation unconformity

| | marine L-stn. Carmel Fmn. | | dolomite Lr.Todilto Fmn. | | gypsum Ur. Todilto Fmn. |
| | eolian sstn. Entrada Fmn. | | fluviatile sstns. conglomerates & siltstones Morrison Fmn. |

Figure 3.4. Reconstruction of Entrada sandstone palaeogeography. Cross-section about 300 ft thick. Modified from Tanner, 1965, Fig. 7. Reproduced from the *Journal of Sedimentary Petrology*, by courtesy of the Society of Economic Paleontologists and Mineralogists. Thin arrows: slope controlled seaward flowing fluvial palaeocurrents. Thick arrows: slope independent eolian palaeocurrents.

piedmont fans to the east. There are no known eolian sands of Upper Permian to Middle Triassic age.

In Late Triassic and Early Jurassic time dune fields developed in inland basins away from the sea. Middle Jurassic eolian sandstones are unknown. In Upper Jurassic time both coastal and interior basin dune fields developed (Fig. 3.5). Tanner (1965) has produced a

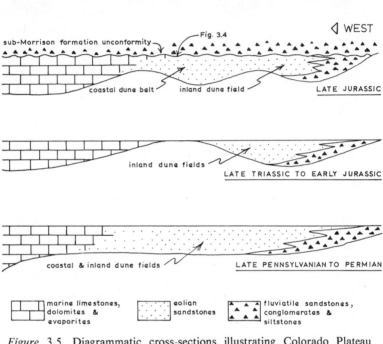

Figure 3.5. Diagrammatic cross-sections illustrating Colorado Plateau eolian sandstone palaeogeographies.

particularly elegant detailed study of the Upper Jurassic palaeogeography of the 'Four Corners' region of the Colorado Plateau. Marine deposits with southwesterly directed longshore palaeocurrents interfinger eastwards with the Entrada sandstone. This was deposited by a complex of southeasterly migrating coastal dunes which, for a time, separated the open sea from the hypersaline deposits of the Todilto Formation lake. Tongues of fluvial sediment within the eolian Entrada sandstone show northwesterly directed cross-bedding indicating the direction of the palaeoslope as the river flowed down to the sea (Fig. 3.4).

GENERAL DISCUSSION OF EOLIAN SEDIMENTS

The Colorado Plateau sandstones just described can be attributed to wind-deposition with considerable confidence. It is not always so easy to recognize eolian sediments for a number of reasons which will now be discussed.

First, unlike many sedimentary facies, eolian deposits have no predictable geometry and no cyclic motif of sub-facies. Lithologically, they are essentially homogenous and with irregular plan and cross-sectional geometries. The only real exception to this is the linear trend of coastal dunes.

Though probably the majority of recognized ancient eolian deposits are quartz sandstones many other lithologies can be wind-deposited. As already seen, cross-bedded gypsum dunes occur in the Jurassic Todilto Formation of the Colorado Plateau. Many Recent inland and coastal salt pans, or sabkhas are rimmed by fields of gypsum dunes, e.g. the White Sands area of New Mexico, and parts of the Trucial Coast respectively. Coastal dune belts are often made of marine-derived carbonate sands complete with a marine fauna sometimes with unfragmented macro-fossils. Pleistocene calcarenites crop out around the shores of many parts of the world today, notably in the Arabian Gulf, southern and eastern Mediterranean, and the Bahamas (Mackenzie, 1964; Sanders and Friedman, 1967, p. 231). These have generated many a discussion on their eolian or marine origin. Not all eolian deposits are of sand grade. The Pleistocene loess silt deposits of North America, Europe, and China are widely interpreted as due to wind action in arid zones around the ice caps.

Many geologists have proposed textural criteria to distinguish wind-blown from water-laid sand. It is widely accepted that eolian sands are better sorted than aqueous ones. This is generally true, but there is no arbitrary dividing limit. The classic eolian Triassic Bunter Sandstones of England have a Trask sorting coefficient of 1·41 (Shotton, 1937). This is not so well-sorted as a beach sand cited by Krumbein and Pettijohn (1938, p. 232) which has a value of 1·22. It is widely accepted that eolian sands are positively skewed, with a tail of fines (Folk, 1966, p. 88). This factor can generally be used to distinguish them from beach sands, but not from fluvial ones (Friedman, 1961, p. 514). Though, as discussed on page 5, textural criteria have been developed to distinguish eolian sands from aqueous ones in Recent sediments it is hard to use these in ancient deposits for technical reasons. Eolian sands are widely believed to be very well rounded and certainly experiments show that wind is much more efficient at rounding quartz sand than is running water (Kuenen, 1960). It is important to note, however, the polycyclic history of the eolian sandstones of the Colorado Plateau, the Permo-Trias of Europe, and the present Sahara. One would expect these sands to be well-rounded and well-sorted, whether they were eolian or not.

Eolian sand grains often show a frosted, pitted surface under the optical microscope and, under the higher powers of the electron microscope, show a variety of characters which can be used to distinguish them from sands subjected to aqueous and glacial action (Krinsley and Funnel, 1965). All these textural features of eolian sands must be carefully interpreted; remembering that dunes may be re-worked by running water, and alluvial fans may dry out and be re-worked by wind. Water-laid deposits may therefore inherit eolian textural parameters and vice versa. For this same reason wind-faceted pebbles (ventifacts, dreikanters) should be interpreted with care.

Sedimentary structures have often been used to distinguish wind-blown from water-laid sediment. As already noted (p. 55) long, low ripples have been recorded from wind-blown sediments and Tanner (1967) states that an R.I. of > 15 can be used to distinguish wind ripples from those due to water, except for swash zone ripples on beaches. This writer has also empirically derived a number of other statistical criteria for distinguishing eolian and aqueously formed ripples. One might have supposed that contorted sand bedding was restricted to aqueous deposits. Unfortunately this is not so. McKee (1966, p. 69) has recorded it in Recent dunes. It occurs in the Colorado Plateau eolian sandstones and is particularly widespread in the Navajo Formation (Stokes, 1961, p. 158).

Cross-bedding with great set heights is generally held to be restricted to eolian sands. However, insufficient is known of Recent marine sand bars to give the maximum set height of water-formed cross-bedding. Therefore no arbitrary limit between eolian and aqueous set heights may be given. Eolian foresets are often no higher than water-laid sets, and only a fraction of the height of the dune in which they occur (see Mckee, 1966, Fig. 8).

Eolian dune foresets are often said to be asymptotically curved towards their bases, and to have steeper inclinations than water-laid foresets. An angle of about 30° is often quoted as critical for distinguishing water-laid from wind-blown foresets. Angles in excess of this figure have been widely recorded from Recent dunes and supposed ancient ones (e.g. McKee, 1966; Mackenzie, 1964; McBride and Hayes, 1963; Laming, 1966; Poole, 1964). On the other hand Shotton (1935) recorded a mean value of only 24° for the Triassic Bunter dune sands of England and Tanner (1965) recorded dips between 21 and 26° as common in the Entrada sandstone of the Colorado Plateau. Care should also be taken in studying foreset inclination data from ancient rocks since it is hard to determine the horizontal datum of the time of their formation. Major bedding surfaces generally represent old erosional slopes of the dune surface, and could seldom have been horizontal (see also Potter and Pettijohn, 1963, pp. 79, 80, and 86).

A further problem of eolian sandstones is the determination of their transport directions. Recent dunes show extremely varied morphologies. Transverse dunes whose foresets consistently dip downwind are generally quantitatively subordinate to barchan (lunate), seif (longitudinal), and stellate dunes. These types have much more complex geometries and correspondingly more varied foreset orientations. Barchan foresets curve through an arc of about 180°, while seif dunes show bipolar foreset dip orientations perpendicular to mean wind direction (McKee and Tibbitts, 1964) (Fig. 3.6). Despite these complexities ancient eolian deposits have produced consistent palaeocurrent trends, though with predictably high scatters of readings. Regional studies have been used in attempts to reconstruct palaeoclimates, particularly the past distribution of high pressure air cells (Shotton, 1956; Opdyke, 1961).

A further problem of eolian palaeocurrents is that they are not slope-controlled like many (but not all) aqueous currents. As we have seen (p. 56) this has been demonstrated in the Entrada sandstone of the Colorado Plateau (Tanner, 1965). In an elegant study of Permotrias desert deposits of southwest England Laming has shown (1966) how sediment was alternately flushed down alluvial fans by floods and blown upslope again by wind.

In conclusion this discussion shows that it is not easy to recognize ancient wind-blown sediments. This is basically because wind and water both transport and deposit sand in ripples and dunes. Many textural and structural criteria have been proposed to distinguish the results of these processes; few, if any, are foolproof. Recognition of an ancient eolian sediment must be based on a critical evaluation of

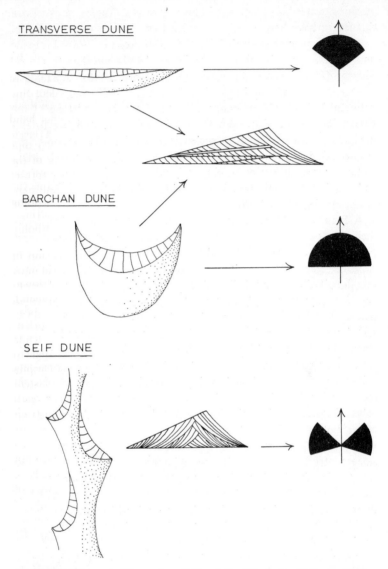

Figure 3.6. Diagrams illustrative of the relationship between dune morphology and cross-bedding orientation shown plotted on azimuths. Wind blows up the page.

all the available data. It may help to distinguish criteria which weigh against an aquatic origin separately from those positively in favour of eolian deposition. Finally it often helps to ask the question: 'If this facies is not eolian, what else could it be?'

ECONOMIC SIGNIFICANCE OF EOLIAN DEPOSITS

Eolian sandstones are potentially of high porosity and permeability because they are typically well-rounded, well-sorted, and generally only lightly cemented. Regional permeability is likely to be good due to absence of shale interbeds. Because of these features eolian sandstones can be important aquifers and hydrocarbon reservoirs. In general, however, eolian sandstones are rated as poor hydrocarbon reservoir prospects. This is because they frequently lie within continental basins far from marine shales which could be potential source rocks. Exceptions to this general rule are found where eolian sandstones overlie pre-existing reservoirs. Renewed tectonic movement and fracturing can allow hydrocarbons to escape from the lower reservoir. They may then migrate upwards and, if there is a suitable cap rock, be retained within the eolian sands. Some of the North Sea gas fields occur within Permo-Triassic dune sands. The source of this gas is thought to be the underlying Upper Carboniferous (Pennsylvanian) Coal Measures. Deep burial of coal beneath the younger sediments may have caused gas to be expelled during devolatization (Bartenstein, 1968).

The distinction of eolian from water-laid sediment can be important where the latter are mineralized and the former are barren. This situation is found in the distribution of Uranium mineralization of the Colorado Plateau (p. 47) and in the Zambian Copper belt (p. 112).

REFERENCES

The account of the Colorado Plateau eolian sandstones was based on:

Baars, D. L., 1961. Permian Blanket Sandstones of the Colorado Plateau. In: Geometry of Sandstone Bodies. (Eds. J. A. Peterson and J. C. Osmond) *Amer. Assoc. Petrol. Geol. Symposium*, pp. 179–207.

Haun, J. D., and Kent, H. C., 1965. Geologic history of Rocky Mountain Region. *Bull. Amer. Assoc. Petrol. Geol.*, **49**, pp. 1781–800.

Lessentine, R. H., 1965. Kaiparowits and Black Mesa Basins: Stratigraphic synthesis. *Bull. Amer. Assoc. Petrol. Geol.*, **49**, pp. 1997–2019.

Opdyke, N. D., 1961. The paleoclimatological significance of desert sandstone. In: *Descriptive Paleoclimatology.* (Ed. A. E. M. Nairn) Interscience, N.Y. pp. 45–59.

Poole, F. G., 1964. Paleowinds in the Western United States. In: *Problems of Paleoclimatology.* (Ed. A. E. M. Nairn) Interscience, N.Y. pp. 390–405.

Stokes, W. L., 1961. Fluvial and eolian sandstone bodies in Colorado Plateau. In: Geometry of Sandstone Bodies. (Eds. J. A. Peterson and J. C. Osmond) *Amer. Assoc. Petrol. Geol. Symposium*, pp. 151–78.

Tanner, W. F., 1965. Upper Jurassic Paleogeography of the Four Corners Region. *J. Sediment. Petrol.*, **35**, pp. 564–74.

Other references cited in this chapter are:

Bagnold, R. A., 1941. *Physics of Blown Sand and Desert Dunes.* London, Methuen & Co. 265 p.

Bartenstein, H., 1968. Present status of the Palaeozoic Palaeogeography of Northern Germany and adjacent parts of Northwest Europe. In: *Geology of Shelf Seas.* (Ed. D. T. Donovan) Oliver & Boyd, Edinburgh and London. pp. 31–54.

Folk, R. L., 1966. A review of grainsize parameters. *Sedimentology*, **6**, pp. 73–93.

Friedman, G. M., 1961. Distinction between dune, beach and river sands from their textural characteristics. *J. Sediment. Petrol.*, **31**, pp. 514–29.

Krinsley, D. H., and Funnel, B. M., 1965. Environmental history of sand grains from the Lower and Middle Pleistocene of Norfolk, England. *Quart. J. Geol. Soc. London.* **121**, pp. 435–61.

Krumbein, W. C., and Pettijohn, F. J., 1938. *Manual of Sedimentary Petrography.* Appleton-Century-Crofts, Inc., N.Y. 549 p.

Kuenen, Ph., 1960. Experimental abrasion: 4. Eolian action. *J. Geol.*, **68**, pp. 427–49.

Laming, D. J. C., 1966. Imbrication, Paleocurrents and other sedimentary features in the Lower New Red Sandstone, Devonshire, England. *J. Sediment. Petrol.*, **36**, pp. 940–57.

Mackenzie, F. T., 1964. Bermuda Pleistocene eolianites and paleowinds. *Sedimentology*, **3**, pp. 52–64.

McBride, E. F., and Hayes, M. O., 1962. Dune cross-bedding on Mustang Island, Texas. *Bull. Amer. Assoc. Petrol. Geol.*, **46**, pp. 546–52.

McKee, E. D., 1966. Structures of dunes at White Sands National Monument, New Mexico (and a comparison with structures of dunes from other selected areas). *Sedimentology*, **7**, No. 1, Sp. Issue, 69 p.

McKee, E. D., and Tibbitts, G. C., 1964. Primary structures of a seif dune and associated deposits in Libya. *J. Sediment. Petrol.*, **34**, pp. 5–17.

Potter, P. E., and Pettijohn, F. J., 1963. *Paleocurrents and Basin Analysis.* Springer-Verlag, Berlin. 296 p.

Sanders, J. E., and Friedman, G. M., 1967. Origin and occurrence of Limestones. In: *Carbonate Rocks*, Part A. (Ed. G. V. Chilingar, H. J. Bissell, and R. W. Fairbridge) Elsevier, Amsterdam. pp. 169–265.

Shotton, F. W., 1937. The Lower Bunter Sandstones of North Worcestershire and East Shropshire. *Geol. Mag.*, **74**, pp. 534–53.

Tanner, W. F., 1967. Ripple mark indices and their uses. *Sedimentology*, **9**, pp. 89–104.

CHAPTER FOUR

LAKE DEPOSITS

INTRODUCTION

Lakes are landlocked bodies of non-marine water. Recent examples may be hundreds of miles long, as are the Canadian Great Lakes, and of widely varying depths. They may be permanent bodies of water or only of short duration. For example, the playa lakes of central Australia last for only weeks or months every few years after torrential rainstorms. Lakes occur in mountainous regions, as in the Alps, and on low-lying continental platforms, like Lake Chad. Their waters range from fresh in temperate and periglacial climates to hypersaline in arid regions. Lakes may be caused by tectonic subsidence and faulting, by glacial erosion, and by damming due to ice, lava, and, recently, man.

Recent lakes have been intensively studied. Most of this work has been concerned with the biology, chemistry, and physics of lake waters. Less attention has been given to bottom sediments but see Hutchinson (1957) and Reeves (1967).

Because of their widely variable environmental settings Recent lacustrine sediments are of many kinds. We must expect ancient lacustrine deposits to show a similar diversity. It is beyond the scope of this book to examine all these different kinds and one case history must suffice. However, the concluding discussion of this chapter attempts to summarize the diverse characteristics of ancient lacustrine deposits.

THE GREEN RIVER FORMATION (EOCENE) ROCKY MOUNTAINS, U.S.A.: DESCRIPTION

At the end of Cretaceous time the Laramide orogeny threw up the Rocky Mountains in a north–south line trending through western central North America. A series of basins developed on the eastern side of these which were infilled by Tertiary continental sediments. During the Eocene period fluviatile and lacustrine deposits were laid down in parts of Wyoming, Colorado, and Utah. Sandstones, silts,

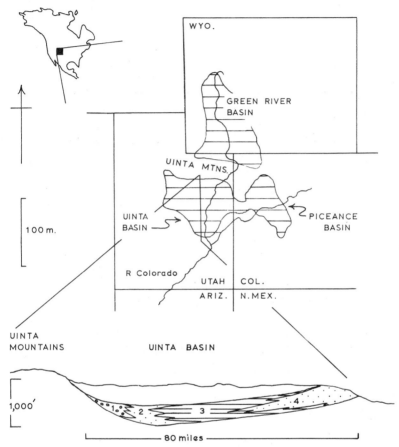

Figure 4.1. Upper: Present approximate areal extent of Green River Formation. Modified from Haun and Kent, 1965. Fig. 23. From the *Bulletin of the American Association of Petroleum Geologists*, by courtesy of the American Association of Petroleum Geologists.

Lower: Cross-section through the Uinta basin showing distribution of Green River facies. 1. Fluvial sand facies. 2. Delta and nearshore sand facies. 3. Oil shale facies. 4. Mud flat deltaic deposits. Based on Osmond, 1965, Fig. 10. From the *Bulletin of the American Association of Petroleum Geologists*, by courtesy of the American Association of Petroleum Geologists.

and coals of the Wasatch Formation are overlain by the lithologically diverse Green River Formation. This is one of the largest and
best documented examples of ancient lacustrine deposits in the
world. Sedimentation of this formation took place essentially in two
areas (Fig. 4.1). The largest of these, Lake Uinta, covered some
23,000 sq. km of northeastern Utah with a depocentre in the Uinta
basin where over 2,200 m of sediment were laid down. This region
was separated by a ridge from the smaller Lake Gosiute with a depocentre in the Green River basin of southwestern Wyoming.

In geometry therefore the Green River Formation is irregular in
plan and thickness, its margins being controlled partly by syn-
depositional topography and tectonics, and partly by subsequent
erosion.

The Green River Formation consists of a wide range of rock types
with complex facies relationships and a correspondingly complex
stratigraphic nomenclature. Two major facies can be recognized;
clastics and carbonates around the margins pass basinward into fine-
grained clastics, carbonates, and evaporites. The marginal facies contains rare conglomerates, cross-bedded and channelled sandstones,
silts, algal dolomites, oolites, and shell beds. This facies interfingers
downwards and laterally into the coal-bearing wholly clastic
Wasatch Formation. The central parts of the basins consist of thick
laminated sequences of marls, fine calcareous sandstones, shales, and
oil shales with saline deposits towards the top. The fine calcareous
sandstones show graded bedding; the oil shales show layered
couplets less than 1 mm thick which are comparable to the varves of
glacial deposits. The lowest part of each couplet consists of very fine
calcareous quartz silt, and the upper part is composed of black
sapropelic material.

Cyclicity is also developed on a larger scale in the marginal deposits, noticeably on the northeastern flank of the Uinta basin. Each
cycle consists of a lower clastic unit which coarsens upwards from
shale to sand and thus to an erosion surface towards the basin margin.
The erosion surfaces are overlain by upward-fining clastic sequences
grading from conglomerate to red shale. This sequence passes upwards and basinwards into oolites and algal carbonates (Fig. 4.2).

The Green River Formation contains an abundant, well-preserved,
and diverse biota. This includes freshwater fish, insects, molluscs,
ostracods, plankton, pollen, calcareous algae, plant remains, and
insects.

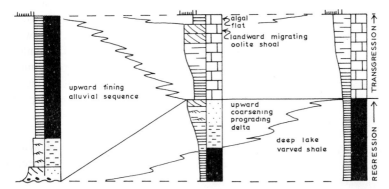

Figure 4.2. Idealized shoreline cycle in the Green River Formation. The fluvial and shoreline cycles (left-hand and central log) vary in thickness from 2–15 m. The deep water cycles (right-hand log) are generally less than 2 m thick. Based on Picard, and High, 1968, Fig. 4. From the *Journal of Sedimentary Petrology*, by courtesy of the Society of Economic Paleontologists and Mineralogists.

In the marginal clastic sediments both cross-bedding and ripple marks have been used to determine palaeocurrents. These indicate sediment transport to and fro up and down local shoreline trends. Palaeocurrent studies have been used to predict the geometries of oil-bearing shoreline sand bodies.

GREEN RIVER FORMATION: DISCUSSION OF ENVIRONMENT

The preponderance of fine-grained laminated sediment shows that this is an aqueous deposit. The absence of a marine fauna indicates that these waters were non-marine, though terminally hypersaline. By definition, therefore, the Green River Formation originated in a lacustrine environment.

With the wealth of data available, however, it is possible to be much more specific about the origin and history of this lake. The absence of cross-bedding and desiccation cracks in the central facies suggests that deposition took place below wave base in waters that never dried out, at least until the terminal hypersaline phase. The varves have been interpreted as annual events, the lower clastic part indicating an influx of land-derived material during the winter rainy period. The sapropelic layers suggest that organic matter from the surface waters rained down on the lake bed during dry summers when there was low runoff and negligible sediment influx from the land. The preservation of this organic matter, together with perfectly preserved fish skeletons and an absence of benthos, suggest that the deeper parts of the lake were anaeorobic and stagnant. Assuming that the varves are annual, knowing their average thickness, and the total thickness of varved sediment, it has been calculated that the lake existed for more than 7,500,000 years. Comparison of the Green River flora with Recent plants indicates that the lake and surrounding hills supported a lush vegetation which grew in a warm humid climate.

The large-scale sedimentary cycles around the margins of the lake have been interpreted as due to transgressions and regressions of the shoreline as the lake waters rose and fell. A drop in level would cause the marginal rivers to downcut to new base levels and carry out the detritus to deposit prograding deltas; hence the upward-coarsening sequences passing basinwards to shale. A rise in lake level would cause the river channels to be choked in their own detritus. Hence the formation of an upward-fining alluvial sequence; the red shale at the top of which suggests the sub-aerial exposure of a floodplain. Simultaneously the drop in clastic supply to the lake would allow near-shore carbonate sedimentation, with oolite banks, shell beds, and calcareous algal mats. These would transgress over the marginal floodplains as the lake rose until the next drop in water level.

It can be seen therefore that the Green River Formation is a good example of an ancient lacustrine deposit. Though some of the more romantic interpretations may be open to question there can be no doubt that these rocks record the past existence of a lake of considerable dimensions.

GENERAL DISCUSSION OF
ANCIENT LAKE DEPOSITS

Basically lacustrine sediments can easily be identified by the combination of low energy aqueous deposits and the absence of a marine fauna. However, as seen in the case of the grey facies of the Torridonian, in PreCambrian sediments with no fauna the distinction between lacustrine and marine sediments may be difficult (p. 32).

The Green River Formation possesses a variety of characteristics which are found in other ancient lakes. In particular note the lithological diversity. Though the majority of lakes are realms of fine-grained sedimentation these may be either argillaceous, calcareous, or evaporitic. Marginal sediments of lakes may be very coarse and conglomeratic, as in the case of the Newark trough (Triassic) lacustrine sediments which pass laterally into fanglomerates (Portland arkose and Newark conglomerate series).

A second feature typical of ancient lakes shown by the Green River Formation is cyclicity. Varves occur not only in this example but, where they were first defined, in Pleistocene glacial lakes of northern Europe (note also that the flora of the Green River Formation demonstrates that varves are not a criterion of glacial climates). Varves also occur in the lacustrine Lockatong Formation (Triassic) of the Newark trough where they occur within large cycles capped by desiccation cracks indicating regular fluctuations and evaporation of the lake waters (Van Houten, 1964). A detailed account of cycles in lacustrine regimes is given by Duff, Hallam, and Walton (1967).

A third feature of some ancient lakes shown by the Green River Formation is the presence of evaporites. These also occur in Recent lakes such as the Dead Sea and the Great Salt Lake of Utah. The source of lacustrine evaporites is often debatable. They may be derived from evaporitic rocks outcropping in the lakes' catchment area. The lake may have formed from a cut-off arm of the sea. Alternatively the salts may have come in the atmosphere from the oceans, being precipitated with rain and concentrated by rivers into ephemeral salt lakes. When attempting to find the source for lacustrine evaporites (where even their lacustrine origin may be in doubt) each case must be judged by its own merits. Thus in the case of the Jurassic Todilto evaporites of the Colorado Plateau it may be significant that they are only a short distance from time equivalent marine strata (p. 58), whereas the Triassic Cheshire salt deposits of northern England, whose origin is in doubt, are far from time equivalent marine limestones and dolomites. Such cases as these need careful observation and interpretation. Even the presence of marine microfossils in evaporites is no proof of a marine origin. Holland (1912) records foraminifera being blown 200 miles inland from the Runn of Kutch to be deposited with evaporites in the Great Sambhar Lake of the Rajputana Desert. This is a useful cautionary note on which to end this discussion.

ECONOMIC ASPECTS

From the preceding data it is apparent that lake sediments may contain oil and gas fields in their marginal sands and oil shales and evaporites in their centres. Important coal fields also occur in ancient fluvio-lacustrine swamps. This type of environmental setting is widespread in Permian coal fields of the continents which are interpreted as once having formed Gondwanaland. These include those of the Karroo System in Zambia, South Africa, and Rhodesia; the Damuda System of India, and equivalent strata in Australia and Antarctica. These coal deposits are all of about the same (Permian) age, they often overlie sediments of possible glacial origin and occur in thick cyclic non-marine clastic sequences. The majority of the coals seem to have originated in swamps on alluvial plains peripheral to lakes (Duff, Hallam, and Walton, 1967, pp. 38–43).

Ancient lake deposits are also important as they can contain economic materials such as china clay (kaolin), iron (generally in the form of limonite), and kiesulguhr (diatomaceous earth).

REFERENCES

The description of the Green River Formation was based on the following references:

Bradley, W. H., 1929. Algal reefs and oolites of the Green River Formation. *U.S. Geol. Surv. Prof. Paper* 154-G, pp. 203–23.

——, 1929. The varves and climates of the Green River Epoch. *U.S. Geol. Surv. Prof. Paper* 158-E, pp. 88–95.

——, 1931. Origin and microfossils of the oil shale of the Green River Formation of Colorado and Utah. *U.S. Geol. Surv. Prof. Paper* 168, 58 p.

——, 1931. Non-glacial marine varves. *Am. J. Sci.*, **222**, pp. 318–30.

——, 1948. Limnology and the Eocene lakes of the Rocky Mountains region. *Bull. Geol. Soc. Amer.*, **59**, pp. 635–48.

Picard, M. D., 1957. Criteria for distinguishing lacustrine and fluvial sediments in Tertiary beds of Uinta Basin, Utah. *J. Sediment. Petrol.*, **27**, pp. 373–7.

——, 1967. Paleocurrents and shoreline orientations in Green River Formation (Eocene), Raven Ridge and Red Wash areas, Northeastern Uinta Basin, Utah. *Bull. Amer. Assoc. Petrol. Geol.*, **51**, pp. 383–92.

Picard, M. D., and High, L. R. Jr, 1968. Sedimentary cycles in the Green River Formation (Eocene), Uinta Basin, Utah. *J. Sediment. Petrol.*, **38**, pp. 378–83.

Other references cited in this chapter were:

Duff, P. McL. D., Hallam, A., and Walton, E. K., 1967. *Cyclic Sedimentation*. Elsevier, Amsterdam. 280 p.

Holland, Sir T. H., 1912. The origin of Desert Salt deposits. *Proc. Lpool. Geol. Soc.*, **11**, pp. 227–50.

Hutchinson, G. E., 1957. *A Treatise on Limnology*. J. Wiley, N.Y. 1015 p.

Reeves, C. C., 1967. *Introduction to Palaeolimnology*. Elsevier, Amsterdam. 228 p.

Van Houten, F. B., 1964. Cyclic lacustrine sedimentation, Upper Triassic Lockatong Formation, central New Jersey and adjacent Pennsylvania. *Kansas Geol. Surv. Bull.*, **169**, pp. 497–531.

CHAPTER FIVE

DELTAS

INTRODUCTION

Shorelines: a general statement

The development of shoreline sedimentary environments and their concomitant facies is a function of many variables, including rate of influx of land-derived sediment, tidal regime, current system, climate, and the relative movements of land and sea. Recent shorelines are sometimes areas of net erosion or net deposition. The former do not

I. LOBATE SHORELINES: DELTAS *Mississippi type*, with radiating 'birdfoot' distributary channels. *Nile type*, radiating channels truncated by an arc of barrier islands.	Carbonate facies encroach shorewards
II. LINEAR SHORELINES: BARRIER ISLAND AND LAGOONAL COMPLEXES *Clastic*, terrigenous sedimentation dominant in the alluvial coastal plain, lagoon, barrier, and open marine environments. *Mixed carbonate: clastic*, terrigenous alluvial plain and lagoon; carbonate barrier and open marine facies. *Carbonate*, limestone open marine and barrier facies. Carbonate facies in lagoonal environment, sometimes dolomitic and evaporitic. Terrigenous coastal plain facies absent or only poorly developed.	Relative increase in land-derived sediment

Table 5.1. A classification of shoreline deposits.

concern us since they leave no sediment in the geological record. The product of depositional shorelines, on the other hand, compose a large percentage of the world's sedimentary cover. The nature of shoreline deposits basically reflects the relative strengths of two processes: the rate of injection into the shoreline system of land-derived sediment; and the ability of marine processes to re-distribute it.

The interplay of these two processes produces a continuous spectrum of shoreline types ranging from the complete dominance of the shoreline by terrestrial sediment, to the other extreme where marine processes dominate over a negligible influx of detritus. With declining terrestrial sediment supply, carbonate sedimentary environments migrate closer and closer inshore.

These concepts can be used as a basis on which to arbitrarily classify this continuous spectrum of shorelines into a number of types, as shown in Table 5.1, page 74.

This chapter describes lobate (deltaic) shorelines. The next three chapters are concerned with clastic, mixed, and carbonate shorelines respectively.

Recent deltas

Herodotus applied the Greek letter delta (Δ) to the triangular area where the distributaries of the Nile deposit their load into the Mediterranean (ibid., 454 B.C.). This term has gained widespread usage by geographers. More recently Lyell (1854) re-defined a delta as 'alluvial land formed by a river at its mouth'. This second definition is perhaps safer for geologists to use since it lays no stress on the deltaic geometry of such an area. Indeed, many Recent river mouths which are accepted as deltaic are far from triangular in shape (see Shirley and Ragsdale, 1966, pp. 234–51). Nevertheless, it is of extreme economic importance to be able to distinguish lobate deltaic shorelines from linear ones. Fortunately this can often be achieved from the study of a single borehole or surface section without acquiring sufficient data to reconstruct the facies geometry. This is because these two types of shorelines deposit distinctive vertical sequences of grain size and sedimentary structures. Deltas are one of the most studied and written about of all Recent environments. A very brief selected list of this vast literature is given at the end of this chapter.

These studies show that deltas form where a river brings more sediment into the sea than can be re-deployed by marine currents. In such a case the river deposits progressively finer sediment in progressively deeper water as its current velocity diminishes seawards. The channel advances seaward between two raised banks of its own deposits, termed levees. Ultimately the channel lays down so much sediment at its mouth that this is higher than the proximal, landward

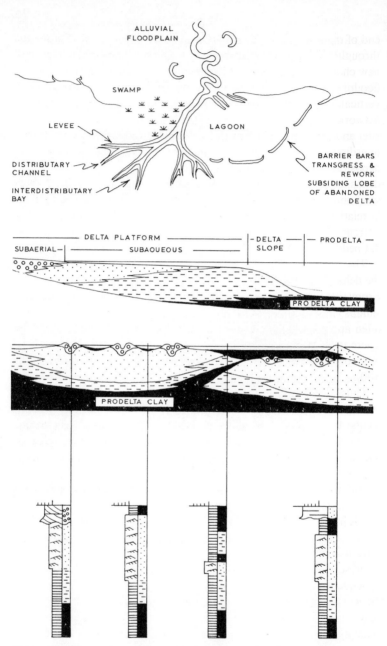

Figure 5.1. Illustrative of the geomorphology and sedimentary facies of a Recent delta. Note the complexity of the vertical sections which may result.

end of the levees. As the channel chokes in its own detritus, it breaks through the levees on its margin. From the crevasse thus formed a new channel flows seaward between two new levees. Thus an alluvial floodplain builds out to the sea, lobate in plan and wedge-shaped in vertical cross-section. The top of the delta consists of a radiating network of distributary channel sands flanked by levee silts. Finer silts and clays are laid down in the flood basins and lagoons of the interdistributary areas. Swamps often form and are the site of peat-deposition. Seaward of the distributary channel mouth, sediment is deposited on the sub-aqueous delta platform. Rippled interlaminated sand and mud are formed, often with bioturbation. If sedimentation is relatively slow, the platform facies may be re-worked by marine currents. The mud may be winnowed out to leave cross-bedded or flat-bedded sands.

The sub-aqueous delta platform is separated from the prodelta by the delta slope. On this surface most of the mud finally comes to rest. From time to time this slope often becomes so steep that the mud slumps and slides down slope to be re-deposited at its foot. Slumping often appears to generate and move down gullies in the delta slope. There is some evidence in the Mississippi and Fraser deltas that this is associated with turbidity currents (Shepard, 1963, pp. 494 and 500).

Thus a vertical section of a delta reveals a fine-grained marine facies, which passes up transitionally into coarser freshwater sediments; these facies boundaries will be diachronous seaward as the delta builds out. Vertical repetition of this sequence may be expected, due to repeated crevassing and delta switching.

In searching for ancient deltas, therefore, we must look for thick clastic sequences showing repeated cycles of upward-coarsening grain size. Each cycle should begin, at the base, with a marine shale which passes up through silts into coarser freshwater channel sands at the top. In plan the channels should show a radiating shoestring pattern and be cut into freshwater shales and coals. Coals may also be present on the top of the channels (Fig. 5.1).

Sequences similar to this occur in rocks of all ages, in all parts of the world. The Carboniferous Period seems to have been a particularly favourable time for delta-formation and the following case history is of this age.

CARBONIFEROUS DELTAIC SEDIMENTATION
OF NORTHERN ENGLAND

At the end of the Devonian Period, a marine transgression took place across northern England. This sea advanced over a tectonically unstable region of actively moving fault-bounded blocks and basins. The sedimentary facies of the Lower Carboniferous are diverse, and their distribution is tectonically controlled. Due to years of study of their fauna, the complex stratigraphy of these rocks is largely resolved. A useful up-to-date account of this work has been given by Rayner (1968).

In general, it is possible to recognize five major sedimentary facies in the Dinantian and Namurian rocks of northern England:

(*i*) *Limestones*, including calcilutites, oolites, and bioclastic calcarenites with marginal reef complexes, were deposited on the relatively stable shelves.

(*ii*) *Black shales* were deposited mainly in the basinal areas.

(*iii*) *The Millstone Grit* facies of coarse pebbly sandstones overlies parts of the shelf limestones and basinal shales.

(*iv*) *Greywacke sandstone and shale* facies, including the Mam Tor Series and Shale Grit of the Central Pennine Trough.

(*v*) *Yoredale* facies: cyclic sequences of limestone, shale, sandstone, and coal occur on the shelf areas above, or time equivalent to the limestone facies previously described.

The last three of these five facies are believed to be essentially of deltaic origin. Two quite different types of delta seem to have operated. Their deposits will now be described and discussed under the arbitrary designations of Type 1 and Type 2, respectively.

Type 1: description

The first type of Carboniferous deltaic sediments to be described is the Yoredale Series. These range from Dinantian to Namurian in age and extend from the Midland Valley of Scotland to the southern edge of the North Pennine Block (Fig. 5.2).

Basically the Yoredale Series consist of eight major cyclothems, each about 100 ft thick, with the general motif: limestone, shale, sandstone, coal. Within this main pattern minor cycles are present,

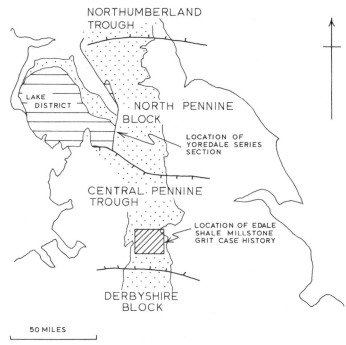

Figure 5.2. Carboniferous tectonic setting of northern England. Outcrop of Dinantian and Namurian rocks stippled; older rocks horizontally ruled; younger rocks blank.

notably thin shale, sandstone, coal sequences towards the top of major sandstones. Figure 5.3 shows two upward-coarsening shale, sandstone sequences between two limestones from the Yoredale Series of Westmoreland.

The limestones of the Yoredale Series have extensive sheet geometries which are relatively easy to correlate laterally in contrast to the terrigenous beds. They are composed of a variety of carbonate rock types ranging from cross-bedded skeletal calcarenites to massive or thinly bedded calcilutites. Silicified limestones and thin chert layers occur especially in the upper part. Cross-bedding and terrigenous sand grains are often typical of the low part of each limestone sequence. Banks of crinoid ossicles and patch reefs of algae, bryozoa, corals, and brachiopods are not uncommon.

The limestones, within each cycle, are generally overlain by shales. These are black pyritic micaceous siltstones with thin siderite concretionary layers. They are laminated throughout and generally unfossiliferous.

The shales pass up gradually with increasing grain size into silty medium- to fine-grained sandstones which are generally laminated or micro-crosslaminated and sometimes burrowed. This sand facies is

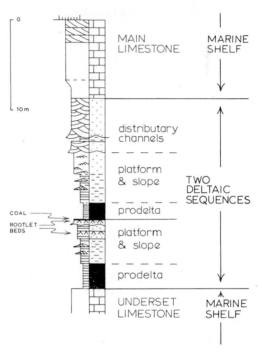

Figure 5.3. Measured section of Yoredale Series, Hell Gill beck, Westmorland (for location see Fig. 5.2). Note how the overall upward-coarsening clastic sequence between the two limestones contains two sub-sequences, only the lower one of which is capped by a coal.

overlain abruptly by coarse cross-bedded sandstones. These occupy large channels, generally of only local extent but sometimes tens of feet deep. These often cut not only into the underlying sandstone, but are also sometimes incised into the shale, or sometimes even the limestone beneath (see for example Moore, 1959, Fig. 8).

The channel sands become finer grained towards the top where they are interbedded with thin coals, carbonaceous shales, and rootlet beds.

Type 1: discussion

The Yoredale Series consist of a series of cycles, each of which shows a progressive change from a marine environment (shown by the limestone fauna) to an upward-coarsening clastic sequence with rootlets at the top which indicate continental conditions. As discussed at the beginning of the chapter these are the features to be expected in a delta.

Considered in more detail the fauna and lithology of the limestones suggest a marine shelf environment with no influx of terrigenous sediment. Lime mud was deposited under low energy conditions, presumably below wave base. The cross-bedded skeletal calcarenites indicate intermittent current action (see Chapter 8 for further data on marine shelf environments). These conditions were terminated by an influx of terrigenous sediment. Initially, as the laminated shales show, this was fine grained and deposited in quiet relatively deep water out of suspension. The vertical increase in grain size and cross-lamination suggest progressive shallowing and increase in current action. Still stronger currents are indicated by the cross-bedded coarse channel sands. Once these were deposited, however, quiet conditions prevailed with peat swamps colonizing the abandoned channel surfaces and inter-channel areas.

Considered in vertical sequence therefore the facies of the Yoredale Series can be attributed to the following environments:

FACIES	ENVIRONMENT	
Coals and rootlet beds	Swamp	
Cross-bedded sands	Alluvial distributary channels	Deltaic
Rippled sands	Sub-aqueous delta platform	
Laminated silts	Prodelta	
Limestones	Marine shelf	

This type of deltaic deposit which built out into open marine carbonate shelf facies is in marked contrast to that which formed where the deltas debouched into the deeper Carboniferous basins.

Type 2: description

The second case history of deltaic sediments to be described from the Carboniferous of Northern England concerns the Millstone Grit and underlying sands and shales. An extremely well-documented example of this sequence is known from the northern edge of the Central Pennine trough. A condensed account of this work follows (see Figs. 5.2 and 5.4).

The sequence begins, at the base, with the Edale Shales which are about 700 ft thick. These are dark grey mudstones often carbonaceous and with disseminated pyrite. The only sedimentary structure which they show is a well-developed and laterally continuous lamination. The shales are largely unfossiliferous but occasionally yield thin shelled bivalves termed *Posidonia* and goniatites. The cephalopods are generally restricted to thin black horizons and are important biostratigraphic index fossils.

The Edale Shales pass up into the Mam Tor Series. These are between 300–400 ft thick, consisting of laterally persistent interbedded sandstones and shales generally less than 2 ft thick. The sandstones are fine-grained greywackes composed of grains of quartz, lithic fragments, and feldspars set in a carbonaceous clay matrix. The base of each sandstone is erosional, with flute and groove marks incised into the shale beneath. Internally the sands show graded bedding and a tendency for each unit to show the following structures arranged from bottom to top in the sequence: massive, flat-bedded, rippled.

The Mam Tor Series pass up into the Shale Grit, which has a thickness of between 400 to 700 ft. This consists largely of greywacke sandstones petrographically similar to those beneath. Shales are rare. The sandstones are thicker bedded however than those of the Mam Tor Series, sometimes with sequences up to a 100 ft thick composed of sand beds 2–10 ft thick. Graded bedding, lamination, and ripple marking are rare. Many of the sands are massive. These sandstones and the rarer shales are cut into and infill a number of spectacular channels up to 50 or more feet deep. These are most common in the upper part of the Shale Grit and also occur in the lower part of the overlying Grindslow Shales. These are between 300–50 ft thick and succeed the Shale Grit with an abrupt base. They are sub-divided into an upper and a lower unit. Their lower part, some 200 ft thick, is an upward-coarsening sequence of laminated mudstones, siltstones, and fine flat-bedded and rippled silty sandstones. Burrows become increasingly common towards the top. The upper part of the Grindslow Shales consists of a heterogenous assemblage of siltstones and fine sandstones which are often

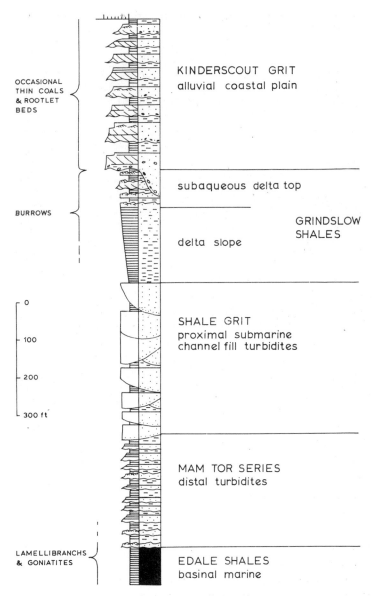

Figure 5.4. Generalized composite section of the Edale Shale : Kinderscout Grit sequence. Location shown in Fig. 5.2. Based on data due to Collinson, Reading, and Walker (see references).

Figure 5.5. Tabular planar cross-bedding in coarse fluviatile sandstone. Kinderscout Grit. 'The Warren', Hathersage. Photo by courtesy of J. D. Collinson.

laminated, rippled, and intensively burrowed. These are pierced by channels of coarser cross-bedded sand.

The Grindslow Shales are overlain by the Kinderscout Grit, the local representative of the Millstone Grit. This consists essentially of over 400 ft of coarse pebbly arkosic sandstones with smaller quantities of shale.

At its base the Kinderscout Grit infills a series of spectacular channels incised into the Grindslow Shales beneath. These are often over 100 ft deep and nearly 1,000 ft wide. They are infilled with very coarse pebbly sandstone, generally massive at the base and flat-bedded or cross-bedded higher up. The overlying part of the Kinderscout Grit consists of fining-upward sequences. These begin, at the base, with an erosion surface overlain by coarse cross-bedded sandstones often with quartz pebbles and shale fragments (Fig. 5.5). The sands become finer grained, laminated, and rippled upwards passing into the laminated grey siltstones with thin coals and rootlet horizons.

Type 2: interpretation

The sequence from the Edale Shales to the Kinderscout Grit shows an overall vertical increase in grain size and a change from marine to continental conditions as shown by the goniatites and rootlet horizons respectively. These two criteria are typically deltaic as discussed at the beginning of the chapter. Considered in more detail, however, there are significant differences between the Edale–Kinderscout sequence and the Recent delta model and the Yoredale Series. In particular the Edale–Kinderscout sequence consists actually of two upward-coarsening sequences. These are of dissimilar facies and cannot be the result of delta switching.

The Edale Shales indicate deposition of mud out of suspension in a low energy environment. The goniatites show that this was marine. The disseminated pyrite indicates anaerobic stagnant reducing conditions. The graded greywackes of the Mam Tor Sandstones and the Grindslow Grit suggest deposition from turbidity currents. The reasons for this diagnosis are not given here. They are discussed in Chapter 10. The deep channels in the upper part of the Shale Grit were probably cut by the turbidity flows which debouched on to the basin floor to deposit the thinner more regularly bedded Mam Tor greywackes. It is possible therefore to designate the Grindslow Grit deposits 'proximal' turbidites which cut and infilled submarine canyons. The Mam Tor sandstones may be termed 'distal' turbidites deposited on fans radiating from the canyon mouths. This association of canyons and turbidite fans is well known from Recent near-shore marine basins such as the San Diego Trough (Hand and Emery, 1964).

The top of the Shale Grit marks the end of the first upward-coarsening sequence. The base of the overlying Grindslow Shales marks the start of the second.

The laminated silts and silty fine sands at the base of the Grindslow Shales suggest deposition out of suspension below wave base. The vertical transition of these deposits into the coarser rippled burrowed and channelled upper part of the Grindslow Shales suggest a progressive shallowing of the environment with deposition from traction currents becoming more and more important.

The deep channels at the base of the overlying Kinderscout Grit clearly herald the advance of an extremely violent current system

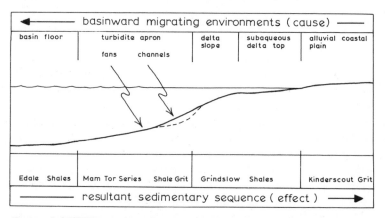

Figure 5.6. Cross-section diagram illustrating the relationship between facies and environments in the Carboniferous rocks of the Central Pennine Trough.

which cut channels deep into the Grindslow Shales and infilled them with coarse sediment. The overlying upward-fining sand:shale cycles suggest deposition from smaller channels which migrated laterally across an alluvial floodplain (p. 23). Coals and rootlets indicate intermittent sub-aerial exposure. Considered together the lower part of the Grindslow Shales can be attributed to deposition on a delta slope. The rippled burrowed silty sands above represent delta platform deposits laid down in interdistributary areas between the cross-bedded sandstone channels. Finally the delta top was buried by the advancing alluvial floodplain.

The Edale Shale–Kinderscout Grit sequence records therefore the gradual infilling of the Central Pennine trough first by turbidites and then by a delta. Palaeocurrent studies show that these deposits came into the area from the North Pennine block, the various facies presumably migrating diachronously southwards (Fig. 5.6).

GENERAL DISCUSSION OF DELTAIC SEDIMENTATION

The two case histories just described illustrate two contrasting types of delta. The Yoredale deltaic deposits built out relatively thin clastic wedges on to shallow marine carbonate shelf deposits.

The Millstone Grit delta on the other hand built out in to a deep marine basin. A steeper delta slope developed therefore and generated aprons of turbidite deposits at its foot. There are other examples of ancient deltaic sediments which can be attributed to the high slope and low slope varieties just defined. High slope deltas, where marine shales pass up through turbidites to channel sands, have been described from the Coaledo Formation (Eocene) of Oregon (Dott, 1966), from Ordovician rocks of the Appalachians (Horowitz, 1966), and from the Westphalian of North Devon (De Raaf, Reading, and Walker, 1965). Low slope deltaic deposits where marine shales (and sometimes limestones) pass through a clean-washed sheet sand facies into alluvial channels are typical of the Pennsylvanian (Upper Carboniferous) of the Illinois basin, U.S.A. The Anvil Rock sandstone is a particularly good example of this type. A lower sheet phase, some 20 ft thick, is laminated or rippled and passes down transitionally into marine shale. Locally, the sheet phase is cut through by coarser cross-bedded channel sands which meander down the palaeoslope (Hopkins, 1958; Potter and Simon, 1961). The Khreim Group (Ordovician–Silurian) of the Tabuk basin of Arabia show a series of deltaic cycles, the lowest of which is a high slope turbidite variant. The overlying cycles suggest relatively low slope conditions. Within each sequence laminated silts pass up through rippled and burrowed interlaminated silts and sands, which underly the upper channel sandstones (Fig. 5.7).

It might be thought from the preceding discussion that such well-documented deposits as deltas were fully understood. This is not true. Many problems remain to be solved, particularly the origin of cyclicity in deltas. This phenomenon is common in ancient deltaic deposits. It has been studied intensively, especially in Carboniferous rocks in which the distribution of deltaic facies seems to be nearly world-wide (Duff, Hallam, and Walton, 1967, p. 81). Studies of

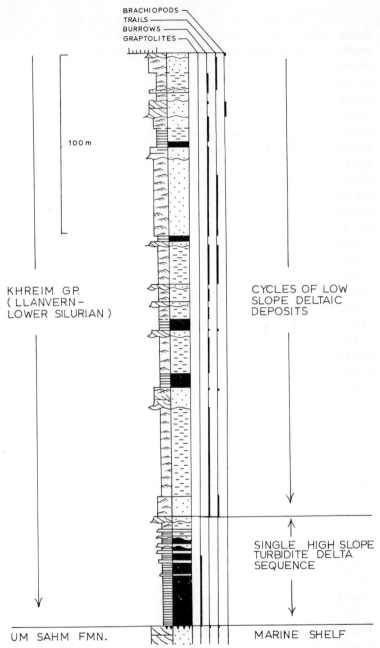

BRACHIOPODS
TRAILS
BURROWS
GRAPHOLITES

100 m

KHREIM G.P.
(LLANVERN–
LOWER SILURIAN)

CYCLES OF LOW
SLOPE DELTAIC
DEPOSITS

SINGLE HIGH SLOPE
TURBIDITE DELTA
SEQUENCE

UM SAHM FMN.

MARINE SHELF

Figure 5.7. Composite measured section of coarsening-upward cycles from the Tabuk basin, Arabia.

Recent deltas suggest that the crevassing of distributary channels provides a built-in cycle generator, no external cause need be invoked (Moore, 1959; Coleman and Gagliano, 1964). Explanations of deltaic cycles which ignore this fact seem improbable (e.g. the tectonic hiccup hypothesis for the Yoredale cycles proposed by Bott and Johnson in 1967). On the other hand, not all cycles in deltaic sediments can be attributed to delta switching. It is possible to correlate individual marine marker bands from Carboniferous deltas as far apart as Ireland and Germany (Ramsbottom and Calver, 1962). Since these extend beyond individual deltas and basins, they can be due neither to crevassing nor sporadic basin subsidence. Eustatic changes are a much more probable explanation for this phenomenon (e.g. Wanless and Shepard, 1936).

It would seem, therefore, that, while deltas can generate their own cycles, external factors such as regular tectonic bumps, climatic and sea-level changes, will also leave their marks on the deltaic pile. In understanding deltaic sequences it is necessary to develop criteria to distinguish the local delta cyclic motif (which must be present) from cycles of external origin which may or may not be present.

There are two useful lines of approach which may help to solve this problem; both employ computers to perform complex statistical gymnastics. The first approach is to study the thickness and regional variation of cyclothems and to see how these relate to other geological variables (e.g. Read and Dean, 1967; 1968).

The second approach uses the computer to simulate deltaic sedimentation; calibration of variables such as sediment input and rate of subsidence being based on Recent examples. (Oertel and Walton, 1967; Bonham-Carter and Sutherland, 1967; Bonham-Carter and Harbaugh, 1968.)

ECONOMIC SIGNIFICANCE OF DELTAIC DEPOSITS

Deltaic sediments are important sources of coal, oil, and gas. Peat-formation is typical in the swamps and marshes of Recent deltaic alluvial plains. The sedimentology of Recent peats and the coals which are their ancient analogues is described in a volume edited by Dapples and Hopkins (1967). In the effective exploitation of coals, it is obviously important to understand their depositional environment. Not all coals are deltaic; as already mentioned some occur in continental basins far from the sea (p. 72). Within the broad frame-work of a delta, coals can form in a number of different situations, each often with a distinct geometry and geological setting. Wanless, Barroffio, and Trescott (1969) have described the various ways in which coal occurs in Pennsylvanian (Upper Carboniferous) rocks in eastern U.S.A. Basically, there are three main situations. First coal may occur in shoestrings trending down the palaeoslope where peat once grew in or on abandoned distributary channels. Second, coals may have shoestring trends parallel to the palaeostrike where they formed in lagoons behind barrier sands. Third, coal seams may have sheet geometries if peat growth was widespread over all the delta after a marine regression. To predict the geometry of a particular coal seam it is obviously important to recognize to which of these three main types it belongs. Detailed mapping of channel sandstones is also important in the exploitation of coal measures. Channels sometimes

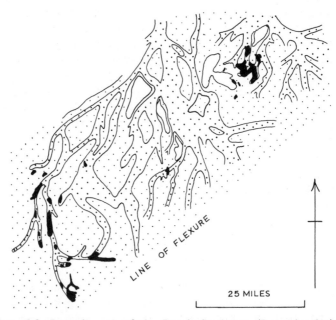

Figure 5.8. Isopach map of the Booch Sandstone (Pennsylvanian) of Oklahoma. Sheet phase (0–20 ft) sparse stipple, channel phase (20–200 ft) dense stipple. See how oil fields (black) are concentrated in the channel sands. Modified from Busch, 1961, Fig. 12. Reproduced by courtesy of the American Association of Petroleum Geologists.

locally downcut and removed coal beds. Furthermore channel sands are often aquifers which should be avoided to prevent flooding of the mines. Channel trends can be predicted from outcrop studies and borehole data used in conjunction with palaeocurrent analysis.

Deltas may also contain significant quantities of oil and gas. This is because they consist of marine shales (potential source rocks) which interfinger up slope with sandstones (potential reservoirs). Obviously it is important to understand the sedimentology of these sands to predict the location and geometry of a hydrocarbon reservoir. In particular, it may be significant to distinguish sub-aqueous delta platform sands from those of the distributary channels. The sheet geometry of the former should make them easy to find. However, since they may be poorer sorted and finer grained than the channel sands, they may have poorer porosity and permeability. Channel sands will be harder to find, however, and, once located, the geometry of a reservoir may be difficult to predict. A classic example of a complex of oil fields located in deltaic distributary channels is shown in Fig. 5.8. A further important point is the distinction between deltaic distributary channels, in which oil pools will trend down slope, and barrier sand reservoirs which are aligned parallel to the shore. This is discussed in the next chapter (p. 106).

REFERENCES

A very brief list of works on Recent deltas, and reviews of Recent and ancient deltaic sedimentation:

Allen, J. R. L., 1965. Late Quaternary Niger Delta, and adjacent areas: Sedimentary environments and lithofacies. *Bull. Amer. Assoc. Petrol. Geol.*, **49**, pp. 547–600.

Fisk, H. N., 1961. Bar finger sands of Mississippi Delta. In: *Geometry of Sandstone Bodies*. (Ed. J. A. Peterson and J. C. Osmond) Amer. Assoc. Petrol. Geol. pp. 29–52.

——, McFarlan, E., Kolb, C. R., and Wilbert, L. J., 1954. Sedimentary Framework of the modern Mississippi Delta. *J. Sediment. Petrol.*, **24**, pp. 76–100.

Moore, D., 1966. Deltaic Sedimentation. *Earth Science Reviews*, **1**, pp. 87–104.

Shepard, F. P., Phleger, F. B., and Van Andel, T. H. (Eds.), 1960. *Recent Sediments Northwest Gulf of Mexico*. Amer. Assoc. Petrol. Geol. 394 p.

Shirley, M. L., and Ragsdale, J. A. (Eds.), *Deltas in their Geologic Framework*. Houston. Geol. Soc. 251 p.

Oomkens, E., 1967. Depositional sequences and sand distribution in a deltaic complex. *Geol. Mijnb.*, **46**, pp. 265–78.

Van Andel, T. H., 1967. The Orinoco Delta. *J. Sediment Petrol.*, **37**, pp. 297–310.

Van Straaten, L. M. J. U., 1960. Some recent advances in deltaic sedimentation. *Lpool. Manchr. Geol. J.*, **2**, pp. 411–43.

The description of Carboniferous deltaic sedimentation of England is based on:

Allen, J. R. L., 1960. The Mam Tor Sandstones: A 'turbidite' facies of the Namurian deltas of Derbyshire, England. *J. Sediment. Petrol.*, **30**, pp. 193–208.

Collinson, J. D., 1966. Antidune bedding in the Namurian of Derbyshire, England. *Geol. Mijnb.*, **45**, pp. 262–4.

——, 1969. The sedimentology of the Grindslow shales and the Kinderscout Grit: A Deltaic Complex in the Namurian of Northern England. *J. Sediment. Petrol.*, **39**, pp. 194–221.

Johnson, G. A. L., 1959. The Carboniferous Stratigraphy of The Roman Wall district in western Northumberland. *Proc. Yorks. Geol. Soc.*, **32**, pp. 83–130.

Moore, D., 1958. Yoredale Series of Upper Wensleydale and adjacent parts of northwest Yorkshire. *Proc. Yorks. Geol. Soc.*, **31**, pp. 91–146.

——, 1959. Role of deltas in the formation of some British Lower Carboniferous cyclothems. *J. Geol.*, **67**, pp. 522–39.

Rayner, D. H., 1967. *The Stratigraphy of the British Isles*. Cambridge University Press.

Reading, H. G., 1964. A review of the factors affecting the sedimentation of the Millstone Grit (Namurian) in the Central Pennines. In: *Deltaic and Shallow Marine Deposits*. (Ed. L. M. J. U. Van Straaten) Elsevier, Amsterdam. pp. 340–6.

Walker, R. G., 1966a. Shale Grit and Grindslow Shales; transition from turbidite to shallow water sediments in the Upper Carboniferous of northern England. *J. Sediment. Petrol.*, **36**, pp. 90–114.

——, 1966b. Deep channels in turbidite-bearing formations. *Bull. Amer. Assoc. Petrol. Geol.*, **50**, pp. 1899–917.

Other references mentioned in this chapter:

Bonham-Carter, G. F., and Harbaugh, J. W., 1968. Simulation of geologic systems: an overview. In: Colloquium on Simulation. (Ed. D. F. Merriam and N. C. Cocke) *Computer Contribution* No. 22. State Geol. Surv. Univ. Kansas. pp. 3–10.

Bonham-Carter, G. F., and Sutherland, A. J., 1967. Diffusion and settling of sediments at river mouths: a computer simulation model. *Trans. Gulf Coast Assoc. Geol. Soc.*, **17**, pp. 326–38.

Bott, M. H. P., and Johnson, G. A. L., 1967. The controlling mechanism of Carboniferous cyclic sedimentation. *Quart. J. Geol. Soc. Lond.*, **122**, pp. 421–41.

Busch, D. A., 1961. Prospecting for stratigraphic traps. In: *The Geometry of Sandstone Bodies*. (Ed. J. A. Peterson and J. C. Osmond) Amer. Assoc. Petrol. Geol. pp. 220–32.

Coleman, J. M., and Gagliano, S. M., 1964. Cyclic sedimentation in the Mississippi River deltaic plain. *Gulf Coast Assoc. Geol. Soc. Trans.*, **14**, pp. 67–80.

Dapples, E. C., and Hopkins, M. E. (Eds.), 1969. Environments of Coal Deposition. *Geol. Soc. Amer.*, Sp. Pap., No. 114.

De Raaf, J. F. M., Reading, H. G., and Walker, R. G., 1964. Cyclic sedimentation in the Lower Westphalian of north Devon. *Sedimentology*, **4**, pp. 1–52.

Dott, R. H., 1966. Eocene Deltaic Sedimentation at Coos Bay, Oregon. *J. Geol.*, **74**, pp. 373–420.

Duff, P. McL. D., Hallam, A., and Walton, E. K., 1967. *Cyclic Sedimentation*. Elsevier, Amsterdam. 280 p.

Hand, B. M., and Emery, K. O., 1964. Turbidites and topography of North end of San Diego Trough, California. *J. Geol.*, **72**, pp. 526–42.

Herodotus, *c.* 430 B.C. *The Histories of Herodotus of Halicarnassus.* Papyrus Publishing Coy. Old Cairo, also 1962. Translated by H. Carter, Oxford Univ. Press. 617 p.

Hopkins, M. E., 1958. Geology and Petrology of the Anvil Rock Sandstone of Southern Illinois. *Ill. State. Geol. Surv. Circ.*, No. 256, 49 p.

Horowitz, D. H., 1966. Evidence for deltaic origin of an Upper Ordovician sequence in the Central Appalachians. In: *Deltas in their Geologic Framework*. (Ed. M. L. Shirley and J. A. Ragsdale) Houston. Geol. Soc. pp. 159–69.

Lyell, Sir Charles, 1854. *Principles of Geology.* 9th Edn. Appleton & Coy, New York.

Oertel, G., and Walton, E. K., 1967. Lessons from a feasibility study for computer models of coal-bearing deltas. *Sedimentology*, **9**, pp. 157–68.

Potter, P. E., and Simon, J. A., 1961. Anvil rock sandstone and channel cutouts of Herrin (No. 6). Coal in West-Central Illinois. *Ill. State Geol. Surv. Circ.*, **34**, 12 p.

Ramsbottom, W. H. C., and Calver, M. A., 1962. *Comptes Rendus 4ᵉ. Congr. Advanc. Etud. Stratigr. Carb.*, pp. 571–6.

Read, W. A., and Dean, J. M., 1967. A quantitative study of a sequence of coal-bearing cycles in the Namurian of central Scotland, 1. *Sedimentology*, **9**, pp. 137–56.

Read, W. A., and Dean, J. M., 1968. A quantitative study of a sequence of coal-bearing cycles in the Namurian of central Scotland, 2. *Sedimentology*, **10**, pp. 121–36.

Shepard, F. P., 1963. Submarine Canyons. In: *The Sea*, vol. III. (Ed. M. N. Hill) Wiley, New York. pp. 480–506.

Wanless, H. R., and Shepard, F. P., 1936. Sea level and climatic changes related to Late Paleozoic cycles. *Bull. Geol. Soc. Amer.*, **47**, pp. 1177–206.

——, Barroffio, J. R., and Trescott, P. C., 1969. Conditions of deposition of Pennsylvanian coal beds. In: Environments of Coal Deposition. (Ed. E. C. Dapples and M. E. Hopkins) *Geol. Soc. Amer.*, *Sp. Pap.*, No. 114, pp. 105–42.

CHAPTER SIX

LINEAR CLASTIC SHORELINES

INTRODUCTION: RECENT LINEAR CLASTIC SHORELINES

Deltas only form where rivers bring more sediment into the sea than can be re-worked by marine current. By their very nature, therefore, deltaic sequences indicate a regression of the shoreline. Where marine currents are strong enough to redistribute land-derived sediment, linear shorelines are formed with bars and beaches running parallel to the coast.

Both deltas and linear shorelines deposit sediment in a wide range of sedimentary environments ranging from continental to marine. Studies of Recent sediments show that both linear and lobate shorelines can form upward-coarsening regressive sequences. Because deltas and linear shorelines both deposit porous sands around marine basins they are important hydrocarbon reservoirs. Since their sand body geometries are quite different, however, it is important to be able to distinguish the two types.

Studies of Recent clastic linear shorelines suggest that four major sedimentary environments can be recognized (e.g. Shepard, Phleger, and Van Andel, 1960; Curray, 1964; Curray and Moore, 1964; Hoyt, 1968; Hoyt, Weimer, and Henry, 1964; Lagaay and Kopstein, 1964; Pannekoek, 1956; Horn, 1965). These consist of two high-energy zones which alternate seawards with two low-energy zones. Basically from land to sea these are: fluviatile coastal plain, lagoonal and tidal flat complex, barrier island, and offshore marine shelf (Fig. 6.1). Each of these four environments deposit facies which can be distinguished from one another by their lithology, sedimentary structures, and biota. These will now be summarized. Further data are given in the references previously mentioned.

The fluviatile coastal plain will deposit alluvium similar to that described in Chapter 2. Since the coastal plain has a gentle gradient the alluvium will be of the meandering river type. Fine-grained flood-plain sediments will dominate over upward-fining channel sand

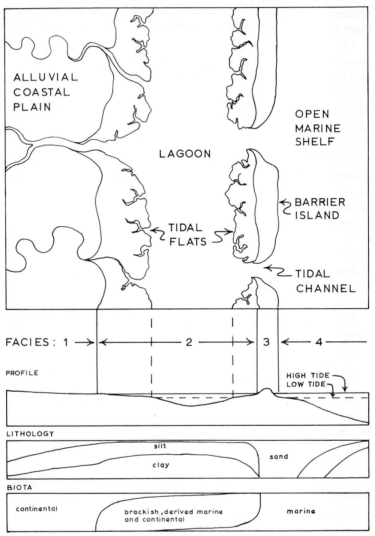

Figure 6.1. Diagram summarizing the geomorphology, sedimentology, and biology of Recent barrier coastlines.

sequences. The biota is continental, with bones, wood, and other plant debris and freshwater invertebrates.

The alluvial coastal plain passes transitionally seawards into swamps, tidal flats, and lagoons. Peat forms in the swamps. Tidal flats deposit delicately interlaminated muds, silts, and very fine sands, often rippled and burrowed. They are cut by meandering gullies in which channel floor lag conglomerates are overlain obliquely by interlaminated fine sediment deposited on prograding point-bars (Van Straaten, 1959; Evans, 1965).

The deposits of lagoons are generally fine-grained too, but depending on their size and depth, may deposit sediment ranging from sand to mud. In regions of low sediment influx carbonate mud can be deposited. Evaporites may form in hypersaline lagoons. The fauna of lagoons is similarly variable depending on the salinity. It may range from freshwater, through brackish (with shell banks) to normal marine or, if restricted and of high salinity, a fauna may be absent. Sedimentary structures of lagoons are similar to those of tidal flats with delicately laminated muds and interlaminated and rippled sand silt and mud. Bioturbation is common.

The lagoon is separated from the open sea by a barrier island complex. This is composed dominantly of well-sorted sand with a fragmented derived marine fauna. A review of Recent barrier deposits (Selley, 1968, p. 431) suggests that the typical internal structure is regular bedding that dips gently seawards. Rare trough and tabular planar cross-bedding dips generally landwards, though bipolar onshore : offshore dips may be recorded (Klein, 1967). The barrier may be no more than an offshore bar exposed only at low tide, or it can form an island with eolian dunes on the crest. Intermittently along its length the barrier may be cut by tidal channels in which cross-bedded sands are deposited (Armstrong Price, 1961; Hoyt and Henry, 1967). To landward the barrier may pass abruptly into the lagoon with the development of washover fans. Alternatively a tidal flat may intervene. Barriers typically pass seawards with decreasing grain size into an offshore zone where mud is deposited below wave base. This will be laminated and contain a marine fauna. Between this and the barrier is a transitional zone of interlaminated sand, silt, and clay with ripples and burrows (Allen, 1967, Fig. 1).

In summary, therefore, a Recent linear shoreline consists of two high-energy zones and two low-energy zones alternating with one another parallel to the coast. In some instances the barrier sand is forced landwards against the alluvial plain. A lagoonal tidal flat complex is then absent. This situation generally occurs on stormy coasts with low input of sediment from the land (Hoyt, 1968).

The actual sedimentary sequence which is deposited from a linear clastic shoreline is a function both of sediment availability and of the rate of rise or fall of the land and sea.

All four facies described above will only be preserved where there is an abundant supply of land-derived sediment. When this happens and the shoreline is stationary all four environments deposit sequences of the four facies side by side. Such static shorelines as these are rare in the geological column because they need a very delicate balance between sedimentation and rising sea level. The Frio sand (Oligocene) of the northwest Gulf of Mexico coast is an example of an ancient static clastic shoreline (Boyd and Dyer, 1966). Where a high influx of sediment is accompanied by a regression of the shoreline (for which the former may be responsible) all four facies build out one above the other seaward. Such regressive sequences are similar to deltas; essentially coarsening upwards from marine shales at the base into continental sands at the top. High sediment influx accompanied by a relative rise in sea level results in a fining-upwards transgressive sequence which is a mirror image of the former type. Regressive and transgressive shorelines with all four facies preserved occur in the Tertiary Gulf Coast Province (e.g. Rainwater, 1964, Figs. 7 and 8) and in Upper Cretaceous rocks of the American Rocky Mountains. The latter example is described and discussed in this chapter.

Linear clastic shorelines with low sediment influx may or may not have the four environmental belts previously defined. The sequence deposited seldom perfectly reflects these. Where the sea transgresses over a land surface starved of sediment an unconformity will pass up, sometimes with a basal conglomerate, into a marine beach sand. This will fine upwards into a marine low-energy facies of shale, or where there is no land-derived sediment, carbonates. This kind of transgressive sequence is seen in the Lower Cambrian rocks of the northwest highlands of Scotland (Phemister, 1960) and in the Glauconitic Sandstone : Chalk Upper Cretaceous transgression of northwest Europe. Transitions are known between this kind of totally marine transgressive sequence and complete fluvial lagoonal : barrier : shelf transgressions. These typically show an unconformity overlain by fluviatile sands which grade up through beach sand into marine

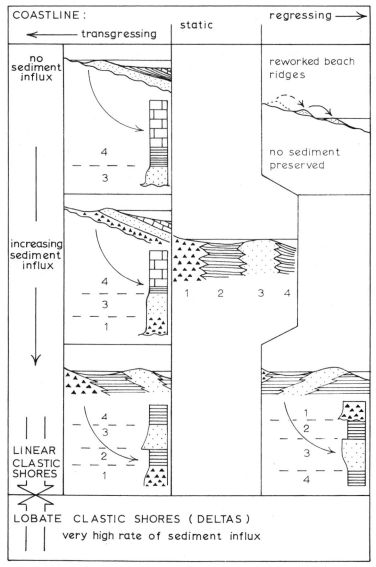

Figure 6.2. Diagrams illustrating the various types of sedimentary sequences which may be deposited by linear clastic shorelines. 1. Continental alluvium. 2. Lagoonal and intertidal deposits. 3. Barrier sands. 4. Open marine shelf facies. For examples of each type of sequence see text.

shales and/or limestones. Such sequences can be seen in the Cretaceous Kurnub sandstone : Ajlun limestone section of Palestine and Jordan (Aharoni, 1966; Bender, 1968) and in the Chicla : Ain Tobi Limestone section of northwest Libya (Hammuda, 1969).

A regression of the sea from a low-lying land surface with no detritus may leave no mark in the geological record other than an unconformity. The actual shoreline may consist of a beach, or barrier beach and lagoon complex. As the sea retreats it will leave the old beach ridges stranded inland. These will be eroded and the sediment carried back to the sea to be re-deposited on the beach face. Net sedimentation above the unconformity is thus generally zero.

Figure 6.2 attempts to summarize the various sequences which may be deposited from linear clastic shorelines.

CRETACEOUS SHORELINES OF THE ROCKY MOUNTAINS, U.S.A.: DESCRIPTION

An easterly thinning clastic prism was deposited from a Cretaceous seaway which stretched across the midwest of North America from Canada to the Gulf of Mexico. Easterly thinning of this sequence is accompanied by a gradual decrease in grain size; coarse conglomerates derived from the rising Rocky Mountains passing eastwards through sandstones to shales. Within these rocks it is possible to recognize three major sedimentary facies which interfinger with one another laterally and are interbedded vertically (Fig. 6.3). These may be listed from west to east as follows:

(*i*) Coal-bearing facies.
(*ii*) Sheet sand facies.
(*iii*) Laminated shale facies.

These will now be described in turn.

Coal-bearing facies

Lithostratigraphically this includes the Menefee Formation of Colorado and New Mexico and the Lance Formation and parts of the Almond and Judith River Formations of Wyoming and Montana.

Geometrically this facies consists of a wedge over 1,000 ft thick in the west which splits up into a number of easterly thinning tongues when traced away from the Rocky Mountains.

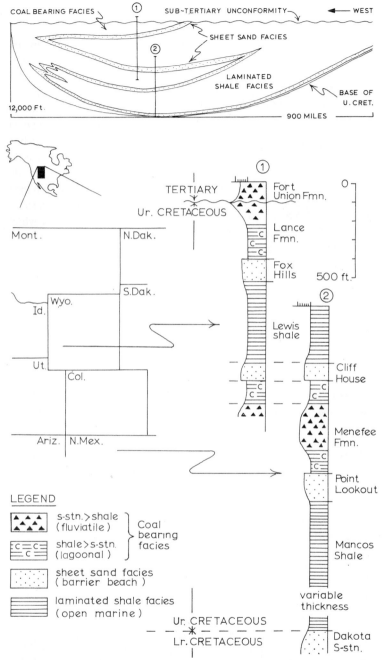

Figure 6.3. Stratigraphic sections of Upper Cretaceous Rocky Mountain shoreline deposits.

In the west this facies is dominantly conglomeratic with subordinate amounts of coarse, poorly sorted sandstones and siltstones. Interbedded lavas and ashes are present, and the conglomerates and sands are often composed of volcanic detritus. These sediments show prominent channelling and cross-bedding.

Traced eastwards grain size diminishes to medium and fine cross-bedded sands with channelled bases interbedded with shales. The shales are laminated and rippled. They contain rare reptile bones and shells of the freshwater lamellibranch *Unio* and the brackish lamellibranchs *Corbula* and *Ostrea*. The oysters form reefs locally. The shales are generally dark in colour and often carbonaceous with fossil plant remains. Locally the shales and sands are interbedded with coals. These are sometimes up to 10 ft thick and have been of commercial significance.

Sheet sand facies

Lithostratigraphically this includes the Fox Hills Formation and the upper part of the Almond and lower part of the Judith River Formations of Wyoming and Montana. In Colorado and New Mexico sands of this facies are represented by the Cliff House and Point Lookout Formations.

Geometrically this facies occurs in sheet-shaped units separating the coal-bearing shales of the previous facies to the west from the laminated shales of the third facies to the east (Fig. 6.3). Individual sand sheets are about 100 ft thick and can be traced for considerable distances both along the palaeoslope (west–east) and the palaeostrike (north–south). In Colorado and New Mexico the sheets contain local thickened benches which are laterally persistent along the palaeostrike for tens of miles. Isolated sand lenticles occur interbedded with laminated shales east of the main development of sheet sands. These have shoestring geometries with north–south trends. Examples include the Eagle sandstone of Montana and the Two Wells sand lentil of New Mexico.

The petrography of the sheet-sand facies is regionally variable, ranging from protoquartzite to feldspathic sand and sub-greywacke. The Gallop sand sheet in Arizona, Colorado, and New Mexico contains local heavy mineral placer deposits rich in ilmenite. Sorting is poor to moderate with considerable quantities of interstitial clay. Texture, fauna, and sedimentary structures show a regular vertical arrangement within any one sand sheet (Fig. 6.4). In the case of a regressive sand this is as follows. The upper contact of the sand with shales of the coal facies is abrupt. In at least one instance the top of the sand is channelled and infilled with an oyster-bearing siltstone (Weimer, 1961, p. 88). The upper part of the sand sequence shows the coarsest grain size and best sorting of the whole sheet. These fine sands are rarely cross-bedded and typically show low angle (5–15°)

Figure 6.4. Regressive Point Lookout barrier sandsheet passing down transitionally into open marine Mancos shale. From Visher, 1965, Fig. 5. Reproduced from the *Bulletin of the American Association of Petroleum Geologists*, by courtesy of the American Association of Petroleum Geologists.

laterally persistent stratification with, generally, easterly dips (westerly in the case of the Eagle Sand shoestring). This facies often contains burrows of *Ophiomorpha* which are comparable to those produced today by the crustacean *Callianassa* on the tidal and subtidal parts of beaches.

These fine flat-bedded sands grade down into very fine silty sands. These are laminated, colour mottled, and sometimes burrowed. They contain the ammonites *Baculites* and *Discoscaphites* together with the lamellibranchs *Inoceramus* and *Pholadomya*. This second unit of the sheet-sand facies grades down into laminated siltstone.

Laminated shale facies

The third facies of the Cretaceous mid-west sediments includes the Lewis, Mancos, Bearpaw, and Pierre Shale Formations. These are best developed to the east where they locally become calcareous and grade into limestones (e.g. the Niobrara Formation). Traced westwards this facies thins and splits up into a number of tongues which interfinger with the sheet-sand facies. Contacts are gradational with a siltstone sequence separating the shales from the very fine silty sands. In some areas the sheet sands are locally absent, and the laminated shales directly overlie the coal-bearing facies.

Lithologically this facies consists of grey laminated claystones with a fauna of ammonites and lamellibranchs similar to that recorded from the lower part of the sheet-sand facies, together with shark teeth.

CRETACEOUS SHORELINES OF THE ROCKY MOUNTAINS, U.S.A.: INTERPRETATION

The conglomerates and coarse cross-bedded channelled sands of the extreme west were clearly deposited in a high-energy environment by fast traction currents. The finer sediments with which they are interbedded eastwards indicate lower energy conditions. The lamellibranchs in the shales suggest deposition in waters whose salinity ranged from fresh to brackish. The thin coal beds indicate intermittent swamp conditions. Considered over all, therefore, the coal-bearing facies seems to have been deposited on a piedmont alluvial plain which passed eastwards down slope into a region of swamps and brackish lagoons.

The sheet-sand facies suggests that the lagoons were restricted to the east by a higher energy environment. The ammonites and lamellibranchs of the sands indicate that they were laid down in or close to a marine environment. The vertical sequence of increasing grain size and increased sorting is typical of Recent barrier beaches (see p. 97), as also is the sequence of sedimentary structures, with the dominance of low-angle stratification in the upper part of each sand sheet. The evidence points, therefore, to a barrier beach environment for this facies. The channels at the top of each sand sequence were probably cut by tidal currents flowing between the open sea to the east and the brackish lagoons to the west.

The laminated shales of the eastern part of the region were clearly laid down in low-energy conditions. Their fauna indicates a marine environment. It seems most probable, therefore, that the laminated shale facies originated below wave base on a marine shelf to the east of the barrier sands.

From the preceding observations and interpretations it can be seen that these Cretaceous sediments provide a good example of a linear clastic shoreline. All four major environments found in Recent linear clastic coasts are represented in both regressive and transgressive phases of the shoreline. The coal-bearing facies represents the alluvial and lagoonal environments, the sheet sands indicate sand barriers, and the marine shale provides evidence of the open sea environment. It is clear from the repeated vertical interbedding and lateral interfingering of all the facies that deposition took place synchronously in all four environments. Several lines of evidence point to fluctuations in the rate of advance and retreat of the shoreline. The local superposition of marine shales directly on the coal-bearing facies shows that sometimes the sea transgressed too fast for barrier sands to form. This may have been caused either by shortages of land-derived sediment, or by extremely rapid rises of sea level or subsidence of the land.

The isolated sand shoestrings within the marine shales also suggest that sometimes the sea advanced so fast that barriers were no sooner formed than they were submerged to form offshore bars and shoals. The location of these was sometimes controlled by palaeohighs on the sea floor. Thus the Eagle Sand shoestring trends along the crest of an anticline. The westerly orientation of the inclined bedding of this body suggests that the shoal migrated landwards.

By contrast, the thick clean sand benches in the sheet-sand facies suggest that from time to time the shoreline was static. High barriers were then thrown up by the sea on which the sand was continually reworked and from which the clay was winnowed.

In conclusion then it can be seen that these Cretaceous sediments provide a good example of a linear clastic shoreline in which rapid deposition allowed the sediments of all four environments to be deposited during both regressions and transgressions of the coast.

GENERAL DISCUSSION OF
LINEAR CLASTIC SHORELINES

As can be seen from the previous case history, clastic shorelines can be easy to identify. This is true not only for those in which all four environments are preserved during regression. It is also easy to identify the passing of a marine transgression where an unconformity is overlain by a marine-sand facies with, or without, an intervening continental sequence (Fig. 6.2).

There is, however, one particular case where confusion of diagnosis can arise. Since this can often be of economic significance it will now be discussed in some detail.

As already mentioned, prograding linear shorelines and deltas both deposit sequences which are broadly comparable. At the base is a marine shale which coarsens upwards into laminated rippled and burrowed silts. These in turn are overlain by coarser sand grade sediment, often with continental fossils and associated with coals and non-marine shales.

Both deltaic and barrier sands often make good hydrocarbon reservoirs. Deltaic distributary sands occur in down slope trending channels. Barrier sands, in contrast, are shoestrings which trend along the palaeostrike. The distinction of the precise environment of an oil-bearing shoreline sand is vital if the geometry of the reservoir is to be predicted accurately.

The confusion is particularly real in areas where barrier sands develop around active deltas, as in the Recent mouth of the Nile. Similarly, abandoned lobes of the Mississippi delta are presently being re-worked by the sea which advances over old delta topsets. Sand barriers and shoals are thrown up and as they migrate landward leave behind a veneer of marine sand on the old delta land surfaces. An ancient analogue of this process can be seen in the Mullaghmore Carboniferous delta complex in the Sligo basin of Ireland. The channel sands at the top of the upward-coarsening deltaic sequence are overlain by a clean calcareous sandstone unit with fragmented marine fossils.

These Recent and ancient examples of associated marine beach sands and delta distributary sands show how closely associated they may be. Fortunately, however, they can generally be distinguished with a few criteria. The lower marine sections are essentially identical. So also is the transitional zone of interlaminated rippled and burrowed silt and very fine sand. It is unlikely, though, that turbidites would be found in a linear shoreline sequence. Relatively slower deposition allows more time for marine currents to re-work the sediment, and gentle sea floor gradients are the rule. Turbidites, however, are not ubiquitous in deltaic shorelines (see Chapter 5). It is

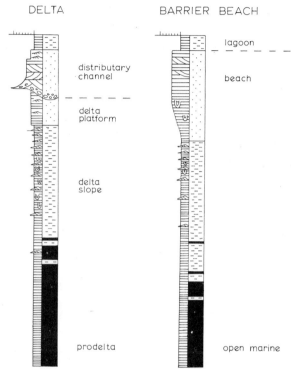

Figure 6.5. Idealized sedimentary sections produced by prograding deltas and barrier sands. Note the close similarity between the lower parts of both sections and the fact that they both coarsen upwards. The main distinguishing features occur in the top sandstone units.

in the upper part of the deltaic and barrier shoreline that differences are present. A deltaic channel sand will generally have an erosional base, often overlain by a thin conglomerate. The succeeding sand will generally be cross-bedded. An upward-fining of grain size may continue into a rippled, very fine sand. Fossils in the channel will contain continental animals and plant debris, often fragmented and showing signs of transportation.

A beach sand on the other hand will have not an erosional but a transitional base passing gradually down into the laminated silty sediment beneath. The sand itself will be typically flat-bedded, with only rare isolated sets of cross-bedding. The fauna will be fragmented, but marine.

The distinguishing features of deltaic and linear shoreline sand sequences are summarized in Fig. 6.5.

It can be seen, therefore, that there are certain criteria which can be used to distinguish bar sands from delta distributary sands. Only one well need be drilled through a sand to determine its origin if it is cored or if the transitional or abrupt base can be picked out on electric logs. Furthermore, if something is known of the palaeogeography of the area it may be possible to predict the orientation of the sand body.

Sometimes, however, nature may make the task of prediction easier. It has already been seen in the case of the Rocky Mountain shoreline that migrating barriers can deposit sheets of sand. Similarly, the top of a delta may consist of a sheet formed of coalesced channels. This can be seen in the deltaic sands of the Tabuk basin of Arabia (p. 87) and in the abandoned La Fourche sub-delta of the Mississippi (Kolb and Van Lopik, 1966, p. 37).

Deltas and linear clastic shorelines are comparable not only because they can both generate upward-coarsening marine shale to continental sand sequences, but also because these are often cyclically repeated. This is shown by the Rocky Mountain Cretaceous sediments. Cyclic linear clastic shoreline sequences are particularly characteristic of the Tertiary Era. They have been described from rocks of this age from as far apart as the Anglo-Paris Basin (Stamp, 1921), the Gulf Coast of Louisiana and Texas (Bornhauser, 1947; Lowman, 1949; Fisher, 1964; MacNeil, 1966), and Sumatra (Dufour, 1951).

Unlike deltas, barrier shorelines contain no sedimentary process which can act as a built-in cycle generator. It is generally agreed, therefore, that such cycles are due to eustatic and/or tectonic causes (Duff, Hallam, and Walton, 1967, p. 187; Tanner, 1968).

ECONOMIC SIGNIFICANCE OF LINEAR CLASTIC SHORELINES

Since linear clastic shorelines deposit clean porous sands around marine basins they are often prolific oil and gas producers. The barrier sands obviously provide the best potential reservoirs with both the marine and the lagoonal shales being potential hydrocarbon source rocks. It is, therefore, very important to be able to recognize bar and beach sands and to make predictions of their geometry and trend. The criteria for distinguishing such sands from deltaic channels has been discussed in the previous section. The significance of this distinction is self-explanatory.

Though at any one time a barrier or beach is a linear environment running parallel to the shore, the geometry of the sand that is actually deposited can be quite varied. There are three main kinds:

(*i*) Regressive sand sheets.
(*ii*) Transgressive sand sheets.
(*iii*) Shoestring barrier sands.

These will now be described in turn.

Regressive sand sheets

The Cretaceous shoreline sediments of the Rocky Mountain region contain good examples of regressive (as well as transgressive) sand sheets. These have established gas reserves in excess of 20 trillion cubic feet with some oil as well. Though some production comes from the continental facies, the bulk is from the marine barrier sands. These sands have sheet geometries, and are, therefore, easy to locate. It is not so easy though to find parts with good reservoir properties. Due to the clay matrix, porosity and permeability values are low. Optimum reservoir characteristics are found where there is secondary fracture porosity and where the sands are particularly well-sorted and clay-free. The second of these features is found in two situations. It occurs at the top of each sand sequence, where it is easy to find, and in the thick sand benches. Since the latter occur in narrow belts often only two or three miles wide, they are not always easy to locate, nor, once found, is their regional trend simple to predict. The Cretaceous shoreline did not extend in a straight line north to south from Canada to the Gulf of Mexico. Like Recent coasts, it had bays, capes, and spits, so that locally the thick barrier sand benches have trends varying from northwest to northeast. Fortunately these strata are gently folded, and due to erosion the barrier sands crop out intermittently at the surface. In such situations the thick porous sands can be located and their trend predicted underground.

Transgressive sand sheets

The geometry of transgressive sands is largely controlled by the shape of the unconformity over which the sea advances. Where this is

a planar surface, the overlying sand will have a sheet geometry.
Though the sand is a sheet, oil reservoirs within it will be linear in
plan, trending parallel to the shoreline. Their up dip limit will
be controlled by the sand pinch out and their down dip extension
will be bounded by the oil : water contact (Fig. 6.6a). Oil reservoirs
occur in basal transgressive sands of both marine and non-marine
origin (e.g. the Lower Cretaceous Cutbank sand of Montana and
the Pliocene Quirequine field of Venezuela respectively (Leverson,
1967, pp. 336–7)).

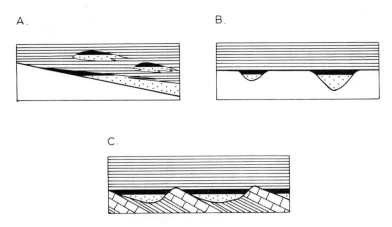

Figure 6.6. Diagrams of some types of stratigraphic oil and gas reservoirs in
clastic linear shoreline deposits. Oil: black; basement: blank; sand: stippled;
shale cap rock: ruled. A. Palaeodip section showing reservoirs in basal
transgressive sand sheet and isolated barrier sand shoestrings. B. Palaeostrike section showing oil accumulation in channel sands. C. Reservoirs in
sands infilling strike valleys cut in softer beds of alternating hard and soft
strata.

More complicated reservoir geometries occur where clastic shorelines transgress over irregular land surfaces. Basal sands, both
marine and non-marine, are then laid down in the low-lying valleys,
but are often absent over the hills. The study of buried land surfaces,
termed palaeogeomorphology, has been discussed from an economic
angle by Martin (1966). Two main kinds of stratigraphic reservoirs
may be found in sands overlying irregular topographies. Old river
channels may trend seawards down the palaeoslope. As the shoreline
advances the fluviatile sands are buried under marine shales. Oil and

gas reservoirs may then form in linear pools parallel to the palaeo-slope (Fig. 6.6b). There are many instances of such reservoirs. Good examples occur in basal Cretaceous sands of Canada (Martin, 1966). A special case of this type of reservoir is where oil accumulates in river terrace sands buried under shale (Conybeare, 1964).

The second type of hydrocarbon accumulation on irregular transgressive surfaces is where sands are deposited in strike valleys perpendicular to the consequent palaeodrainage surface. These form where the rocks beneath the unconformity consisted of interbedded, gently dipping hard (scarp-forming), and soft (valley-forming) strata (Fig. 6.6c). Oil reservoirs in strike valley sands occur in the basal Cretaceous of Alberta and Saskatchewan (Martin, 1966) and in basal Pennsylvanian Cherokee sands of Kansas and Oklahoma (Busch, 1961).

Obviously a considerable understanding not only of sedimentology but also of geomorphology is needed to predict the location of hydrocarbon reservoirs in the basal sands of transgressive clastic shorelines.

Shoestring barrier sands

Shoestring sands of clastic shorelines are of two main types. One occurs as barriers separating marine from non-marine shale. The second type are shoestrings entirely enclosed in marine shale. The origin of such sands has been discussed in the light of data from Recent shelf sea sand bars by Off (1963). Prolific oil production comes from Tertiary barrier shoestring sands of the Gulf Coast province of Texas and Louisiana. Though many of these barriers form transgressive and regressive sheets (Rainwater, 1964, Figs. 7 and 8), shoestrings also occur separating marine from non-marine shale. The Frio sand (Oligocene) is a good example, being up to 5,000 ft thick and 25 miles wide. Oil production extends for several hundred miles along the palaeostrike (Burke, 1958; Boyd and Dyer, 1966). Similar examples occur in the Middle Vicksburg (Lower Oligocene) of the same region. Here barrier sand shoestrings lie to the northeast and southwest of a deltaic complex separating lagoonal shales to the northwest from open marine shales to the southeast. These shoestrings are about 60 ft thick and 3 miles wide (Gregory, 1964).

There are many well-documented cases of oil and gas production from shoestring sands entirely enclosed within marine shale. Particularly good examples are found in Pennsylvanian sediments of Oklahoma and Kansas. Individual bars are two or more miles in width and some 50 ft thick. In plan the shoestrings are often arranged *en echelon*, though trends, both linear and curved, can be traced from upwards of 50 miles (Busch, 1961; Leverson, 1967, Figs. 7–14). Within the Enid embayment of the Anadarko basin of Oklahoma oil accumulations in concentric marine shoestrings are associated with reservoirs in basinward trending channel sands (Withrow, 1968).

Similar fields in offshore bars occur in the Silurian Clinton Sands and Late Devonian : Early Mississippian of Pennsylvania and West Virginia (Potter and Pettijohn, 1963, Figs. 9.3 and 9.4).

Linear clastic shorelines are not only of economic importance because of their ability to trap oil and gas. They may contain coals formed in the swamps and lagoons behind barriers, as in the Cretaceous Mesaverde Formation of the Rocky Mountains. Carboniferous coal measures, though largely deltaic, also contain coals formed behind barrier sands (p. 90). Beach sands can contain winnowed valuable detrital minerals, such as the Recent ilmenite-rich beach sands of India, Ceylon, and New Zealand (Deer, Howie, and Zussman, 1962, p. 31). Ilmenite-rich Cretaceous beach deposits of the Gallop Sandstone in New Mexico are an ancient analogue (Murphy, 1956).

The copper belt of Zambia is another example of a linear clastic shoreline of considerable economic significance. Here a Pre-Cambrian sea advanced across a rugged granite land surface with fluviatile and eolian sands. Pyrite, chalcopyrite, bornite, and chalcocite occur in both the fluviatile sands and in overlying marine shales, sandstones, and dolomites. Mineralization is principally developed in bays and drowned valleys of the shoreline. Regardless of whether the minerals formed at the time of deposition of the sediment, or later, sedimentological studies have an important part to play in unravelling the palaeogeography and environment of such deposits, thus helping to locate ore bodies and to predict their geometry (Mendelsohn, 1961; Garlick, 1969).

In conclusion it can be seen that the study of linear clastic shorelines can be very profitable.

REFERENCES

The account of the Cretaceous shoreline east of the Rocky Mountains was based on:

Hollenshead, C. T., and Pritchard, R. L., 1961. Geometry of producing Mesaverde Sandstones, San Juan basin. In: *The Geometry of Sandstone Bodies*. (Ed. J. A. Peterson and J. C. Osmond) Amer. Assoc. Petrol. Geol. pp. 98–118.

Miller, D. N., Barlow, J. A., and Haun, J. D., 1965. Stratigraphy and petroleum potential of latest Cretaceous rocks, Bighorn basin, Wyoming. *Bull. Amer. Assoc. Petrol. Geol.*, **49**, pp. 277–85.

Pike, W. S., 1947. Intertonguing marine and non-marine Upper Cretaceous deposits of New Mexico, Arizona, and Southwest Colorado. *Geol. Soc. Amer. Mem. No. 24.*

Shelton, J. W., 1965. Trend and genesis of lowermost sandstone unit of Eagle Sandstone at Billings, Montana. *Bull. Amer. Assoc. Petrol. Geol.*, **49**, pp. 1385–97.

Viele, G. W., and Harris, F. G., 1965. Montana Group Stratigraphy, Lewis and Clark County, Montana. *Bull. Amer. Assoc. Petrol. Geol.*, **49**, pp. 379–417.

Weimer, R. J., 1961. Spatial dimensions of Upper Cretaceous Sandstones, Rocky Mountain Area. In: *Geometry of Sandstone Bodies*. (Ed. J. A. Peterson and J. C. Osmond) Amer. Assoc. Petrol. Geol. pp. 82–97.

——, 1966. Patrick Draw Field, Wyoming. *Bull. Amer. Assoc. Petrol. Geol.*, **50**, pp. 2150–75.

——, and Haun, J. D., 1960. Cretaceous stratigraphy, Rocky Mountain Region, U.S.A. *Int. Geol. Congr. Norden*, pt. 12, pp. 178–84.

Other references cited in this chapter:

Aharoni, E., 1966. Oil and gas prospects of Kurnub Group (Lower Cretaceous) in Southern Israel. *Bull. Amer. Assoc. Petrol. Geol.*, **50**, pp. 2388–403.

Allen, J. R. L., 1967. Depth indicators of clastic sequences. In: Depth indicators in marine sedimentary environments. (Ed. A. Hallam) *Marine Geology*, Sp. Issue, **5**, No. 5/6, pp. 429–46.

Armstrong-Price, W., 1963. Patterns of flow and channelling in tidal inlets. *J. Sediment. Petrol.*, **33**, pp. 279–90.

Bender, F., 1968. *Zur geologie von Jordanien*. Beitr. Reg. Geol. Erde. Bd. 7, Berlin.

Bornhauser, M., 1947. Marine sedimentary cycles of Tertiary in Mississippi Embayment and Central Gulf coast area. *Bull. Amer. Assoc. Petrol. Geol.*, **31**, pp. 696–712.

Boyd, D. R., and Dyer, B. F., 1966. Frio barrier bar system of South Texas. *Bull. Amer. Assoc. Petrol. Geol.*, **50**, pp. 170–8.

Burke, R. A., 1958. Summary of oil occurrence in Anahuac and Frio formations of Texas and Louisiana. *Bull. Amer. Assoc. Petrol. Geol.*, **42**, pp. 2935–50.

Busch, D. A., 1961. Prospecting for stratigraphic traps. In: *Geometry of Sandstone Bodies*. (Ed. J. A. Peterson and J. C. Osmond) Amer. Assoc. Petrol. Geol. pp. 220–32.

Conybeare, C. E. B., 1964. Oil accumulation in alluvial stratigraphic traps. *Australasian Oil and Gas. Jl.* August 1964, 5 p.

Curray, J. R., 1964. Transgressions and Regressions. In: *Marine Geology*. (Ed. R. L. Miller) Macmillan. pp. 175–203.

Curray, J. R., and Moore, D. G., 1964. Holocene regressive littoral sand, Costa de Nayarit, Mexico. In: *Deltaic and Shallow Marine Deposits*. (Ed. L. M. J. U. Van Straaten) Elsevier, Amsterdam. pp. 76–82.

Deer, W. A., Howie, R. A., and Zussman, J., 1962. *Rock-forming Minerals. Vol. 5, Non-Silicates*. Wiley, N.Y. 370 p.

Duff, P. L. McL. D., Hallam, A., and Walton, E. K., 1967. *Cyclic Sedimentation*. Elsevier, Amsterdam. 280 p.

Dufour, J., 1951. Facies shift and isochronous correlation. *Proc. 3rd World Petrol. Cong.* Netherlands. Section 1, pp. 428–37.

Evans, G., 1965. Intertidal flat sediments and their environments of deposition in the Wash. *Quart. Jl. Geol. Soc. Lond.*, **121**, pp. 209–45.

Fisher, W. L., 1964. Sedimentary patterns in Eocene cyclic deposits, northern Gulf Coast region. *Kansas Geol. Surv. Bull.* 169, pp. 151–70.

Garlick, W. G., 1969. Special features and sedimentary facies of stratiform sulphide deposits in arenites. In: *Sedimentary Ores Ancient and Modern (Revised)*. (Ed. C. H. James) Sp. Pub. No. 1, Geol. Dept. Leicester Univ., U.K. pp. 107–69.

Gregory, J. L., 1966. A Lower Oligocene delta in the subsurface of southeastern Texas. In: *Deltas*. (Ed. M. L. Shirley and J. A. Ragsdale) Houston Geol. Soc., pp. 213–28.

Hammuda, O. S., 1969. Jurassic and Lower Cretaceous Rocks of Central Jabal Nefusa, Northwestern Libya. *Petrol. Explor. Soc. Libya Guidebook*. 74 p.

Horn, D., 1965. Zur geologischen Entwicklung der Sudlichen Schleimundung im Holozan. *Meyniana*, **15**, pp. 42–58.

Hoyt, J. H., 1968. Genesis of sedimentary deposits along coasts of submergence. *Rept. 23, Int. Geol. Cong. Prague*, p. 231.

——, and Henry, V. J., 1967. Influence of island migration on barrier island sedimentation. *Bull. Geol. Soc. Amer.*, **78**, pp. 77–86.

Hoyt, J. H., and Weimer, R. J., 1963. Comparison of modern and ancient beaches, Central Georgia Coast. *Bull. Amer. Assoc. Petrol. Geol.*, **47**, pp. 529–31.

——, ——, and Henry, V. J., 1964. Late Pleistocene and Recent sedimentation, central Georgia Coast, U.S.A. In: *Deltaic and Shallow Marine Deposits.* (Ed. L. M. J. U. Van Straaten) Elsevier, Amsterdam. pp. 170–6.

Klein, G. de V., 1967. Paleocurrent analysis in relation to modern marine sediment dispersal patterns. *Bull. Amer. Assoc. Petrol. Geol.*, **51**, pp. 366–82.

Kolb, C. R., and Van Lopik, J. R., 1966. Depositional environments of the Mississippi River deltaic plain, southeastern Louisiana. In: *Deltas.* (Ed. M. L. Shirley and J. A. Ragsdale) Houston Geol. Soc., pp. 17–62.

Lagaay, R., and Kopstein, F. P. H. W., 1964. Typical features of a fluviomarine offlap sequence. In: *Deltaic and Shallow Marine Deposits.* (Ed. L. M. J. U. Van Straaten) Elsevier, Amsterdam. pp. 216–26.

Leverson, A. I., 1967. *Geology of Petroleum.* Freeman & Co., San Francisco. 724 p.

Lowman, S. A., 1949. Sedimentary facies in Gulf Coast. *Bull. Amer. Assoc. Petrol. Geol.*, **33**, pp. 1039–97.

MacNeil, F. S., 1966. Middle Tertiary Sedimentary Regimen of Gulf Coast Region. *Bull. Amer. Assoc. Petrol. Geol.*, **50**, pp. 2344–65.

Martin, R., 1966. Paleogeomorphology. *Bull. Amer. Assoc. Petrol. Geol.*, pp. 2277–311.

Mendelsohn, F., 1961. *The Geology of the Northern Rhodesian Copperbelt.* Macdonald, London.

Murphy, J. F., 1956. Preliminary report on Titanium-bearing Sandstone in the San Juan basin and Adjacent Areas in Arizona, Colorado, and New Mexico. *U.S. Geol. Surv. Open File Report*, No. 385.

Off, T., 1963. Rhythmic linear sand bodies caused by tidal currents. *Bull. Amer. Assoc. Petrol. Geol.*, **47**, pp. 324–41.

Oomkens, E., 1967. Depositional sequences and sand distribution in a deltaic complex. *Geol. Mijnb.*, **46**, pp. 265–78.

Pannekoek, A. J., 1956. *Geological History of the Netherlands.* Govt. Printing Office, The Hague. 147 p.

Phemister, J., 1960. *Scotland: The Northern Highlands.* Brit. Reg. Geol. H.M.S.O., London.

Potter, P. E., and Pettijohn, F. J., 1963. *Paleocurrents and Basin Analysis.* Springer-Verlag, Berlin. 296 p.

Rainwater, E. H., 1966. The Geologic Importance of deltas. In: *Deltas.* (Ed. M. L. Shirley and J. A. Ragsdale) Houston Geol. Soc., pp. 1–16.

Selley, R. C., 1968. Nearshore marine and continental sediments of the Sirte Basin, Libya. *Quart. Jl. Geol. Soc. Lond.*, **124**, p. 419–460.

Shelton, J. W., 1965. Trend and genesis of lowermost sandstone unit of Eagle Sandstone at Billings, Montana. *Bull. Amer. Assoc. Petrol. Geol.,* **49**, pp. 1385–97.

Shepard, F. P., Phleger, F. B., and Van Andel, T. H. (Eds.), 1960. Recent sediments, northwest Gulf of Mexico. *Amer. Assoc. Petrol. Geol.* 394 p.

Stamp, L. D., 1921. On cycles of sedimentation in the Eocene strata of the Anglo-Franco-Belgian Basin. *Geol. Mag.,* **58**, pp. 108–14, 146–57, and 194–200.

Tanner, W. F. (Ed.), 1968. Tertiary Sea-level Fluctuations. *Palaeogeography, Palaeoclimatology, Palaeoecology,* Sp. Issue. 178 p.

Weimer, R. J., and Hoyt, J. H., 1964. Burrows of *Callianassa major* SAY, geologic indicators of littoral and shallow neritic environments. *Jl. Pal.,* **38**, pp. 761–7.

Withrow, P. C., 1968. Depositional environments of Pennsylvanian Red fork sandstone in northeastern Anadarko Basin, Oklahoma. *Bull. Amer. Assoc. Petrol. Geol.,* **52**, pp. 1638–54.

Van Straaten, L. M. J. U., 1959. Minor structures of some recent littoral and neritic sediments. *Geol. Mijnb.,* **21**, pp. 197–216.

CHAPTER SEVEN

MIXED CLASTIC: CARBONATE SHORELINES

INTRODUCTION

Mixed carbonate : clastic shorelines are defined as those in which carbonate deposition occurs so close to the land that it contributes, not just to the open sea facies, but also to the shoreline deposits themselves (see p. 74).

These conditions can be brought about by three factors acting singly or in concert. Low input of terrigenous sediment to the shoreline may be due to low runoff or, if the hinterland is low-lying, low sediment availability. Thirdly, if the shoreline itself has a very gentle seaward gradient it will have an extremely broad tidal zone and an extremely wide development of the facies belts paralleling the shore. In such instances, terrigenous sediment may be dumped around river mouths in estuarine tidal flats. There may be insufficient current action to re-work these deposits and carry sand out to the barrier zone. In this case, onshore currents may pile up bars of carbonate sand far to seaward of the river mouths.

A particularly good example of a mixed carbonate : clastic shoreline occurs in Miocene rocks of the Sirte basin, Libya. All three of the factors outlined above seem to have operated in this instance. This case history will now be described.

A LIBYAN MIOCENE SHORELINE: DESCRIPTION AND DISCUSSION

The general tectonic setting and history of the Sirte basin of Libya, in which this shoreline occurs, is described in the next chapter (p. 143). It will only be summarized here. Briefly, the Sirte basin developed in Upper Cretaceous time as a southerly extension from the Tethys geosyncline by block-faulting and subsidence of a part of the Sahara Shield.

By the Miocene period the embayment had largely been infilled by carbonates and shales. Through late Tertiary time a gradual regression of the sea was interrupted by minor marine transgressions. One of these in Lower Miocene time (Aquitanian–Burdigalian) deposited the shoreline sediments now to be described. These have been studied in the region of Marada oasis and the Jebel Zelten (Figs. 7.1 and 7.2).

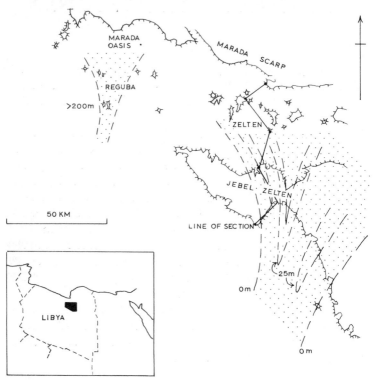

Figure 7.1. Location map of the Miocene shoreline of Marada Oasis and the Jebel Zelten, Libya. Approximate areal extent of major calcareous sand complexes shown stippled. These are aligned along two Miocene synclinal axes which previously, and subsequently, were positive trends.

In the western part of the region, an unconformity between Miocene and Oligocene sediments crops out at the surface. This seems to have been subjected to intensive sub-aerial weathering. The Oligocene limestones immediately beneath the unconformity are extensively oxidized, replaced by gypsum, and penetrated by the roots of trees which grew on the pre-Miocene land surface. The

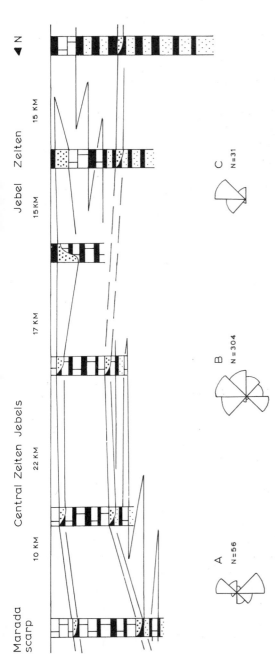

Figure 7.2. Diagrammatic measured sections indicating facies changes across the Miocene shoreline between the Marada scarp and the Jebel Zelten. Datum is post-Marada erosion surface. Right hand section 180 m thick. Key: open marine and barrier limestones: blocks; lagoonal and intertidal shales and sands: black; fluviatile sands: sparse stipple; estuarine calcareous sandstones: dense stipple. Grain size and frequency of cross-bedding in limestones increases over the crest of the anticline in the Central Zelten Jebels. Azimuth diagrams show cross-bedding dip orientations North vertical. Left: limestones show bipolar pattern possibly due to tidal currents; southerly directed major mode indicates dominance of onshore transport. Centre: calcareous sandstones show bipolar pattern with major mode directed offshore, probably due to deposition by tidal currents in estuarine channels. Right: unimodal seawards orientation of cross-bedding shown by fluviatile sands.

succeeding Miocene shoreline deposits, some 200 m thick, are over-lain by a later Miocene formation. The contact with this is uncon-formable in the Jebel Zelten, but becomes transitional when traced northwards towards the centre of the basin. This contact has a present-day northerly dip of about 1 : 1,000. Used as a datum, this surface shows that the underlying Miocene sediments were pre-viously folded into a gentle east–west trending anticline with a cul-mination over the Zelten oil field.

The Miocene shoreline deposits are composed of a diversity of repeatedly interbedded facies which may be listed as follows:

(*i*) Detrital limestone facies.
(*ii*) Laminated shale facies.
(*iii*) Interlaminated sand and shale facies.
(*iv*) Cross-bedded sand and shale facies.
(*v*) Calcareous sand-stone channel facies.

Predominate in north (seawards)

Predominate in south (landwards)

North–south radiating shoestring complexes

Each facies will now be described and interpreted in turn.

Detrital limestone facies

Description
The detrital limestones compose nearly the complete 200 m section exposed in the north. Traced southwards they die out due to inter-fingering with shales and sands. Petrographically these are poorly sorted medium- and coarse-grained packstones, composed of a framework of bioclastic debris and pellets with interstitial micrite and sparite. They still have a high intergranular porosity. Finer calcarenites, calcilutites, and calcirudites are also present. There are considerable regional variations in the grain size of this facies; coarser calcarenites being concentrated over the crest of the Zelten anticline, while finer carbonate sands and muds are present to north and south.

The coarser-grained limestones are flat-bedded or cross-bedded in isolated sets about a metre thick. In plan view it can be seen that both tabular planar and trough foresets are present, often on a vast scale. Individual planar foresets can be traced along strike for over 100 m.

Troughs are up to 50 m wide (Fig. 7.3). The orientation of these structures indicates deposition from alternately onshore (south) and offshore (north) flowing currents, though with a dominance of onshore transport. The finer limestones are generally massive, and often bioturbated. At many points limestone beds are cross-cut by large channels, several metres deep. These are infilled by calcarenites similar to those in which they are cut. The channel margins and floors are often lined with re-worked, bored limestone cobbles and boulders up to a metre in diameter.

Figure 7.3. Wind-eroded surface showing large-scale cross-stratification in offshore bar detrital limestone facies. From Selley, 1968, Plate 21(a), by courtesy of the Geological Society of London.

The limestones are largely composed of a diverse biota in all stages of preservation from entire and obviously *in situ*, to highly comminuted. This includes calcareous algae, bryozoa, corals, lamellibranchs (such as oysters and scallops), gastropods, echinoids, foraminifera (including miliolids and peneroplids), and a diverse suite of trace fossils (including *Ophiomorpha*).

Interpretation
Detailed comparison of the fossils with Recent forms indicate that they grew in a variety of habitats ranging in depth from 0–50 m,

from fully marine to slightly brackish salinities and from low- to high-energy bottom conditions.

The coarse bioclastic cross-bedded limestones were probably deposited in turbulent conditions. Comparison with Recent carbonate sands suggests that they originated from migrating offshore bars and shoals. Recent examples of these deposit foresets on their steep slopes and sub-horizontal bedding on their gentle backslopes (see McKee and Sterrett, 1961, p. 26, Fig. 5). The troughs and channels cross-cutting the shoals may be analogous to those which tidal currents cut through Recent carbonate barriers (e.g. Jindrich, 1969). This interpretation is supported by their bipolar palaeocurrents (Fig. 7.2).

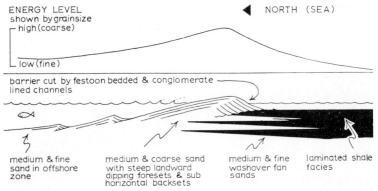

Figure 7.4. Illustrative of the presumed origin of offshore bars in the detrital limestone facies. From Selley, 1968, Fig. 6, by courtesy of the Geological Society of London.

Steep channel margins and intraformational limestone conglomerates testify to penecontemporaneous diagenesis comparable to the formation of 'beachrock' by the sub-aerial exposure of Recent carbonate beaches (e.g. Ginsburg, 1953).

Considered over all, therefore, it seems most probable that the coarser limestones were deposited by shoreward migrating offshore bars which, intermittently exposed above sea level, became barrier islands.

The finer-grained, more generally massive and burrowed calcarenites point to deposition in less turbulent conditions largely below wave base, and perhaps in depths as great as that suggested by the fossils (Fig. 7.4).

Laminated shale facies

Description

Beds of shale 1 or 2 m thick occur across the whole region, interbedded with limestones in the north and sands in the south. They are
grey or green in colour, often highly calcareous and contain thin
white calcilutite bands. Internally, the shales are laminated throughout and occasionally rippled and burrowed.

Figure 7.5. Oyster shell bank with valves in vertical growth position. From
Selley, 1968, Plate 21(b), by courtesy of the Geological Society of London.

Within the shales, and separating them from limestones beneath, it
is common to find oyster reefs. These are often up to a metre thick
and composed of the *in situ* shells of long thin oysters, oriented
vertically with their umbos pointed downwards (Fig. 7.5). Apart
from the oysters, this facies contains rare bryozoa and calcareous
algae and plant debris.

Interpretation

The lamination and fine grain size of this facies indicate deposition out of suspension in a low-energy environment. The fossils suggest a range of salinity from normal marine to brackish. These conditions could be fulfilled either in relatively deep water below wave base, or in shallow water sheltered from the open sea. The brackish element of the fauna and the way in which the shales interfinger seaward with carbonate shoals suggest that the latter alternative is the correct one. The laminated shale facies may, therefore, be attributed to a lagoonal environment.

Interlaminated sand and shale facies

Description

The third facies of this Miocene shoreline occurs in lenticular units two or three metres thick, interbedded with all the other facies. It is erratically distributed through the region, being most common in the Jebel Zelten and lower parts of the section exposed to the north. It consists of sands and interlaminated sand and shales, with rare thin lignites and rootlet beds. The sands are fine-grained, well-sorted, and argillaceous. Internally they are massive, laminated, or micro-crosslaminated. Associated with the sands are delicately inter-laminated very fine sands, silts, and clays. These are typically rippled and highly burrowed with vertical, often U-shaped, sand-filled tubes attributable to the ichnogenus *Diplocraterion* (Fig. 7.6).

Figure 7.6. Interlaminated very fine sand and shale with *Diplocraterion* burrows. From Selley, 1968, Plate 22(b), by courtesy of the Geological Society of London.

Figure 7.7. Fining upward tidal-creek sequence. (1) basal erosion surface, (2) channel-lag conglomerate, (3) cross-bedded coarse-medium sandstone of channel bar, (4) cross-laminated fine sand of point bar, (5) laminated siltstone channel fill. From Selley, 1968, Plate 23, by courtesy of the Geological Society of London.

Cross-cutting these sediments are large curvaceous channels up to 20 m deep and 80 m wide. The channels are infilled by sediments showing a regular sequence of sedimentary structures and a vertical decrease of grain size (Fig. 7.7). The scoured channel floor is overlain by a conglomerate which grades up into several metres of cross-bedded calcareous sand. This is overlain by interlaminated rippled

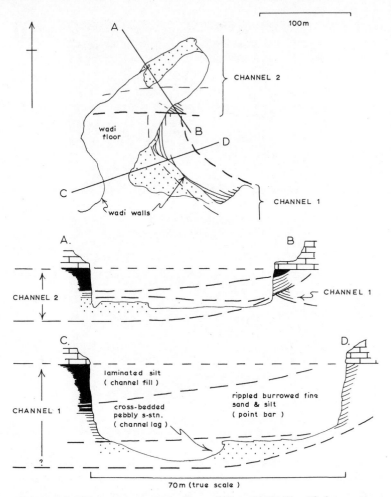

Figure 7.8. Tidal creek channels. From Selley, 1968, Fig. 10, by courtesy of the Geological Society of London.

and burrowed very fine sand, silt, and clay beds which dip obliquely off the channel walls and are succeeded transitionally by horizontally laminated shale (Fig. 7.8).

Interpretation

The over-all fine grain size of this facies indicates deposition in a relatively low-energy environment. Ripples of sand and interlamination of sand, silt, and clay point to sedimentation from gentle currents of pulsating velocity which alternately caused sand ripples to migrate and then halted to allow clay to settle out of suspension. The resultant bedding type and the associated intensive burrowing are closely comparable to Recent tidal flat deposits such as those described from the North Sea coasts (Van Straaten, 1954; Evans, 1965). Likewise, the morphology and sediments of the channels are analogous to those formed by the tidal gullies which drain Recent mud flats (e.g. Van Straaten, 1954, Fig. 4). Further evidence of the shallow environment of this facies is provided by the lignites and rootlet horizons which may have originated in salt marshes on the landward side of the tidal flats. Rare desiccation cracks also testify to the shallow origin of this facies.

Cross-bedded sand and shale facies

Description

This is the most southerly of the four facies belts which are aligned subparallel to the Sirte basin shore. It is best developed in the south scarp of the Jebel Zelten and also occurs, interbedded with the previously described facies, at the base of jebels exposed northwards as far as the Marada escarpment.

This facies is divisible into three interbedded sub-facies. Two-thirds of it is composed of poorly sorted pale yellow unconsolidated sands. These range in grain size from coarse to fine. The coarser sands occur in sequences 5–6 m thick with erosional channelled bases, often veneered by thin quartz pebble conglomerates. Internally they are both tabular planar and trough cross-bedded with set heights of 30–40 cm arranged in vertically grouped cosets. The orientations of foresets and trough axes indicate deposition from unidirectional northerly flowing currents (Fig. 7.2). The finer sands are argillaceous, massive, and, sometimes, flat-bedded.

The second sub-facies, interbedded with the first, consists of laminated shales which occur both as sheets and infilling abandoned channels. They can be distinguished from the laminated shale facies to the north since they are not calcareous and swell and fall apart in water. This phenomenon suggests that they are largely composed of montmorillonite.

The third sub-facies consists of thin sequences, only a few centimetres thick, of lignite, sphaerosideritic limestone, and ferruginized sand pierced throughout by rootlets.

The cross-bedded sand and shale facies contains an abundant, diverse, and well-preserved vertebrate fauna. This includes bones of terrestrial mammals such as ancestral elephants, camels, giraffes, antelopes, and carnivores, together with aquatic forms such as crocodiles, turtles, and fish. These bones occur within the sands together with transported tree trunks. The interbedded shales contain plant debris and the continental gastropod *Hydrobia*.

Interpretation
This facies differs from the three previously described in the dominance of terrigenous sediment and continental fossils. These facts, together with the sedimentary structures and northerly directed palaeocurrents, suggest deposition in a fluviatile environment. (See Chapter 2 for the diagnostic features of alluvium.)

Accordingly, the sand sub-facies can be attributed to deposition within river channels, while the shales and fine sands probably originated on levees and floodplains. The channel-fill shales are abandoned ox-bow lake deposits. The thin limestones, ferruginous layers, lignites, and rootlet beds represent old soil horizons which perhaps formed on the levees and low-lying flood basins between channels. The large amounts of shale imply that the alluvium was due to sinuous meandering rivers rather than braided ones. The vertebrate fauna suggests a savannah climate.

In conclusion, it can be seen that the cross-bedded sand and shale facies originated in a low-lying alluvial coastal plain which, seawards, merged imperceptibly into the tidal flats to the north.

Calcareous sandstone channel facies

Description
The four facies belts paralleling the Sirte basin just described are locally cross-cut by northerly trending calcareous sandstone channels of the fifth and last facies.

These can arbitrarily be sub-divided into two types: small isolated channels of sandy limestone, and large radiating channel complexes of calcareous sandstone.

Channels of the first type are concentrated near the base of the southern scarp of the Jebel Zelten, where they are interbedded with the fluviatile deposits. These also occur at the base of jebels around the Zelten oil field and can be traced at the same level as far north as the Marada scarp. Channels of this type are about 10 m deep and 300 m wide. In some areas they are well-exposed where they have been exhumed from the softer sands and shales with which they are interbedded (Fig. 7.9). These channels are composed of poorly sorted medium- and coarse-grained sandy limestones with fragments of marine shells mixed with quartz sand and micrite matrix. They are floored by intraformational bored limestone pebble conglomerates. The sands are cross-bedded in a wide variety of scales and types, and,

Figure 7.9. Meandering estuarine channel of resistant sandy limestone, exhumed from soft fluvial sands and shales, now crops out as a long sinuous flat-topped jebel. From Selley, 1968, Plate 24(a), by courtesy of the Geological Society of London.

less commonly, massive, flat-bedded, or rippled. Burrowing is present, especially in the finer sediment towards the top of each channel which is often intensively burrowed and sometimes shot through with plant rootlets.

The major channels occur at two points and, due to the good exposure, can be isopached (Fig. 7.1). At Reguba this facies is about 200 m thick. Near the crest of the south scarp of the Jebel Zelten is a sheet of calcareous sandstone, lenticular in an east–west direction, about 25 km wide and 30 m deep. On the north scarp of the Jebel Zelten this has split up into a series of discrete channels which can be traced north to the upper part of the sections in jebels around the

Zelten oil field. One or two sandy channels occur at the same level in the Marada scarp.

Petrographically this facies consists of coarse and very coarse, sometimes pebbly, calcareous sandstones. These are arranged in a series of coalesced channels infilled with various kinds of large-scale trough and planar cross-bedding. These sometimes show penecontemporaneous deformation attributable to quicksand movement. Apart from bioturbation, this facies contains no fossils *in situ*. There are fragments of marine shellfish, bones, teeth, and wood.

Palaeocurrents determined from cross-bedding are bipolar with the major mode pointing northwards (Fig. 7.2). Around the Jebel Zelten palaeocurrents plotted at outcrop show a regionally radiating pattern.

Interpretation

Clearly this facies was deposited from fast-flowing currents confined to channels. The palaeocurrents and the mixture of marine carbonate sediment and shells with terrigenous sand, bones, and wood indicate to and fro current movement. It seems highly probable, therefore, that this facies originated in estuaries subject to strong tidal currents.

The location of the two major channel complexes at Reguba and Jebel Zelten may not be a matter of chance. They both trend along north–south pre-Miocene palaeohighs which host several major oil fields (Fig. 8.3). This suggests that negative movement occurred along these two trends in Miocene time, favouring the development of estuaries where they cross-cut the shoreline.

General discussion of the Miocene shoreline of the Sirte basin

In conclusion, it can be seen that this case history provides a good example of a mixed carbonate : clastic shoreline. From north to south relatively deep water carbonates pass up slope into coarser cross-bedded shell sands, deposited on shoals and bars. These interfinger southwards with fine-grained terrigenous muds and sands of lagoonal and tidal flat origin. In turn these pass landwards into an alluvial coastal plain facies. Locally, this shoreline was interrupted by two major estuaries which supplied coarse sand to the shoreline (Fig. 7.10). Currents were seldom strong enough to carry this detritus out to the bar zone. Subtle syn-sedimentary movement seems to have controlled the distribution of facies. The concentration of coarse grain size and cross-bedding in the limestones over the east–west trending crest of the Zelten anticline suggests that it may have been a Miocene palaeohigh on which the barrier carbonates were deposited. Likewise the two major estuaries seem to have been located along northerly subsiding axes.

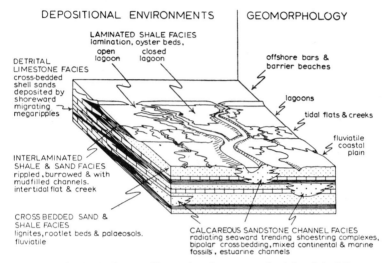

DEPOSITIONAL ENVIRONMENTS | GEOMORPHOLOGY

LAMINATED SHALE FACIES
lamination, oyster beds,
open   closed
lagoon   lagoon

DETRITAL
LIMESTONE FACIES
cross-bedded
shell sands
deposited by
shoreward
migrating
megaripples

INTERLAMINATED
SHALE & SAND FACIES
rippled, burrowed & with
mudfilled channels.
intertidal flat & creek

CROSS BEDDED SAND &
SHALE FACIES
lignites, rootlet beds & palaeosols.
fluviatile

offshore bars &
barrier beaches

lagoons

tidal flats & creeks

fluviatile
coastal
plain

CALCAREOUS SANDSTONE CHANNEL FACIES
radiating seaward trending shoestring complexes,
bipolar crossbedding, mixed continental & marine
fossils, estuarine channels

Figure 7.10. Block diagram illustrating the supposed origin of the Miocene shoreline of the Sirte basin, Libya. From Selley, 1968, Fig. 17, by courtesy of the Geological Society of London.

Palaeocurrent analysis shows that the carbonate sands were deposited by predominantly up slope shoreward-flowing currents, thus coming to rest in waters shallower than that in which they formed. In contrast, quartz sand was carried off the Sahara shield to be deposited in the alluvial plain and tidal flats. Only the finest fraction was transported as far as the lagoons. Mixing of the land-derived quartz sand and marine carbonate detritus occurred only in the estuarine channels, perhaps due to tidal currents.

GENERAL DISCUSSION AND
ECONOMIC ASPECTS

Shorelines where barrier carbonates are juxtaposed with continental terrigenous facies are transitional between clastic shores where carbonate sedimentation, if present, is restricted to the offshore zone, and carbonate shores where terrigenous sediment is negligible. These are described in the previous and subsequent chapters respectively.

Apart from the Libyan Miocene case, other mixed shorelines occur in the Permian rocks of West Texas (Chapter 9). Here shoal calcarenites pass landwards through lagoonal deposits into continental red beds and evaporites. These are overlain by reef limestones with similar shoreward facies changes.

Indeed, the combinations of conditions which favour the occurrence of mixed carbonate : clastic shores (i.e. low influx of terrigenous sediment and aridity) are particularly favourable for reef growth. Perhaps reefs are more characteristic of mixed shorelines than are carbonate sand banks.

These shorelines are of considerable economic significance. This need not be discussed here. The importance of alluvial deposits has already been described on pages 47 and 109. The economic aspects of carbonate bars and reefs are discussed on pages 150 and 176 respectively.

REFERENCES

The account of the Libyan Miocene shoreline was based on:

Doust, H., 1968. *Palaeoenvironmental Studies in the Miocene* (*Libya, Australia*). Vol. I. Unpublished Ph.D. Thesis. University of London. 254 p.

Savage, R. J. G., and White, M. E., 1965. Two mammal faunas from the early Tertiary of Central Libya. *Proc. Geol. Soc. Lond.* No. 1623, pp. 89–91.

Selley, R. C., 1966. The Miocene rocks of Marada and the Jebel Zelten: a study of shoreline sedimentation. *Petrol. Explor. Soc. Libya.* 30 p.

——, 1967. Paleocurrents and sediment transport in the Sirte basin, Libya. *J. Geol.*, **75**, pp. 215–23.

——, 1968. Facies profile and other new methods of graphic data presentation: application in a quantitative study of Libyan Tertiary shoreline deposits. *J. Sediment. Petrol.*, **38**, pp. 363–72.

——, 1968. Nearshore marine and continental sediments of the Sirte basin, Libya. *Quart. Jl. Geol. Soc. Lond.*, **124**, pp. 419–60.

——, 1972. Structural control of Miocene sedimentation in the Sirte basin. In: *The Geology of Libya*. (C. Gray Ed.) The University of Libya. pp. 99–106.

Other references listed in this chapter were:

Evans, G., 1965. Intertidal flat sediments and their environments of deposition in the Wash. *Quart. Jl. Geol. Soc. Lond.*, **121**, pp. 209–45.

Ginsburg, R. N., 1953. Beachrock in south Florida. *J. Sediment. Petrol.*, **23**, pp. 85–92.

Jindrich, V., 1969. Recent carbonate sedimentation by tidal channels in the Lower Florida keys. *J. sediment. Petrol.*, **39**, pp. 531–53.

McKee, E. D., and Sterrett, T. S., 1961. Laboratory experiments on form and structure of longshore bars and beaches. In: *Geometry of Sandstone Bodies.* (Ed. J. A. Peterson and J. C. Osmond) Amer. Assoc. Petrol. Geol., pp. 13–28.

Van Straaten, L. M. J. U., 1953. Composition and structure of Recent marine sediments in the Netherlands. *Leid. Geol. Meded.*, **19**, pp. 1–110.

CHAPTER EIGHT

CARBONATE SHORELINES AND SHELF DEPOSITS

INTRODUCTION: A GENERAL THEORY OF SHELF SEA SEDIMENTATION

The preceding three chapters described shoreline sediments. Where sediment input from the land is so great that marine currents cannot re-distribute it, deltas form. Where marine currents are competent to re-distribute the sediment linear bar and barrier coasts form. With declining sediment input carbonate shoals can develop offshore from alluvial coastal plains. This chapter is concerned with shorelines where the amount of land-derived sediment is so small that carbonate shoreline facies develop. These typically occur on broad tectonic shelves where the hinterland is so low (and sometimes arid) that no clastic detritus is available. The carbonate shoreline grades imperceptibly into an open marine shelf where calcareous or argillaceous deposits form.

There are many areas in the world where relatively thin (generally less than 2,000 ft) limestone sequences cover thousands of square miles with easily correlatable layer-cake stratigraphies. Palaeontology indicates that these deposits were formed in relatively shallow open marine conditions. These deposits, though they may at present occupy tectonic basins, must have been deposited on vast shelves with only the gentlest of bottom slopes.

A general theory of carbonate shelf sedimentation was proposed by Irwin (1965) based on a study of Upper Palaeozoic carbonates of the Williston basin, North America. This case history is described in detail later in this chapter. First Irwin's epeiric sea model will be described and discussed.

Within limestone shelf sequences, such as those of the Williston basin, it is possible to define three major sedimentary facies which pass laterally into one another from the centre of the basin to its margin in the following manner:

(*i*) *Calcilutites:* grading via biomicrites with unfragmented fossils into (*ii*) *Grainstones* (skeletal and oolitic): grading via packstones into

(*iii*) *Pelletoidal limestones, microcrystalline dolomites and evaporites.*

Irwin explained these facies relationships thus: If one considers a marine shelf, it consists of two parallel horizontal surfaces of sea level and wave base. These transect the sloping sea floor shorewards (Fig. 8.1). In the deeper part of the basin, below wave base, fine-grained laminated mud will settle out of suspension. The fauna will generally be preserved *in situ* and unfragmented. These conditions may extend over thousands of square miles. Irwin termed this the 'X' zone. Shorewards, where wave base impinges on the sea floor,

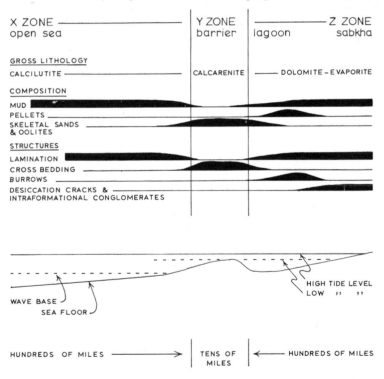

Figure 8.1. Origin of carbonate shelf and shoreline facies. Modified from Irwin, 1965, Fig. 3. From the *Bulletin of the American Association of Petroleum Geologists*, by courtesy of the American Association of Petroleum Geologists.

turbulent bottom conditions will prevail. Lime mud will be winnowed away and carbonate shells fragmented to skeletal sands. Oolites may form. These deposits are piled up into shoals and bars which can be compared with Recent carbonate banks such as those of the Bahamas (Illing, 1954; Newell, Purdy, and Imbrie, 1960; Purdie, 1963). Irwin terms these clean-washed carbonate sands the 'Y' zone. In contrast to the vast areas of the 'X' zone this facies will occur in a linear belt parallel to the coast and only tens of miles wide.

Shoreward of these high-energy barrier sands the sea bed continues its gentle landward climb. In these restricted low-energy waters lagoonal conditions prevail. Recent carbonate lagoons are well-documented in the references cited above (see also p. 155). They are characterized by skeletal and faecal pellet sands which, since they are deposited in quieter conditions than those of the barrier, are micritic (packstones and wackestones). These grade landwards into tidal flats of laminated, sometimes burrowed, carbonate muds where, if salinities are high, dolomites and evaporites may form. There has been considerable speculation whether these are precipitated directly on the floors of lagoons or whether they are due to penecontemporaneous replacement of carbonate mud. In Recent arid shorelines, such as those of the Baja California and the Trucial Coast of the Arabian Gulf, wide salt flats (sabkhas) are developed. In these, evaporite minerals form at the present time (Shearman, 1963 and 1966; Holser, 1966; Kinsman, 1969). It seems that, as the sun beats down, capillarity draws lagoonal brines into the pore spaces of the sabkha carbonates. Here, as the fluids evaporate and concentrate, they replace the host sediment to form dolomite, gypsum, anhydrite, halite, and other evaporite minerals. This facies of pelletal limestones, dolomites, and evaporites Irwin termed the 'Z' zone.

This then is the interpretation which may be put forward as a general explanation of carbonate shelf shoreline sedimentation.

Mississippian deposits of the North American Williston basin will now be described and discussed since they are the type example of this suite of rocks.

MISSISSIPPIAN (LOWER CARBONIFEROUS) DEPOSITS OF THE WILLISTON BASIN, NORTH AMERICA: DESCRIPTION

The Williston basin covers many thousands of square miles of the states of Montana and the Dakotas of the U.S.A. and the provinces of

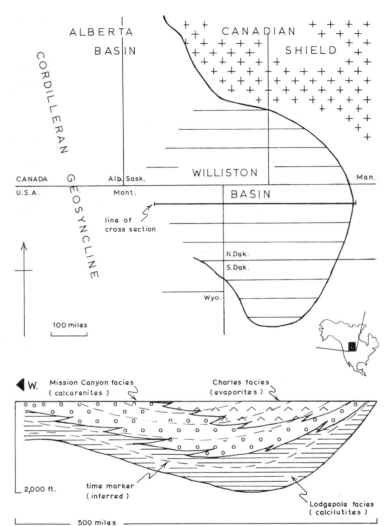

Figure 8.2. Upper: map of the Mississippian (Lower Carboniferous) deposits of the Williston basin. Modified from Carlson and Anderson, 1965, Fig. 1. From the *Bulletin of the American Association of Petroleum Geologists* by courtesy of the American Association of Petroleum Geologists. Lower: cross-section of Mississippian (Lower Carboniferous) deposits of the Williston basin. Modified from Smith *et al.* 1958, Fig. 3, by courtesy of the American Association of Petroleum Geologists.

Alberta, Saskatchewan, and Manitoba in Canada. It contains over 15,000 ft of sediments of all geological periods. These rocks are largely of shallow-water origin and have suffered only the gentlest tectonic deformation, basinward dips being locally interrupted by anticlinal flexures. In Upper Palaeozoic time carbonate sedimentation took place from the Williston basin northwards through the Alberta and Mackenzie basins to the Arctic. These deposits consist largely of shales, limestones (sometimes reefal), dolomites, and evaporites. Oil-bearing Devonian reefs are described in the next chapter. Petroleum is also present in Mississippian carbonates of the southeastern edge of the Williston basin in Canada and northern U.S.A. Intensive study of these beds led to the concept of X, Y, and Z zones stated in the previous section. These rocks will now be described.

The Mississippian section of the Williston basin is over 2,000 ft thick with a depocentre in the western part of North Dakota. Stratigraphical nomenclature is complex and regionally variable. For the purposes of this account the system proposed by Carlson and Anderson (1965, p. 1340) is used (Fig. 8.2). A series of stratigraphic intervals are defined by thin persistent evaporites and clastics which can be correlated regionally in sub-surface by gamma-ray peaks on well logs. The marker horizons which define these intervals are believed to approximate to time planes. Three facies diachronously cross-cut these from top to bottom as follows:

(*i*) *Charles facies:* cyclic evaporites.

(*ii*) *Mission Canyon facies:* bioclastic limestones, dolomites, and oolites.

(*iii*) *Lodgepole facies:* thin-bedded argillaceous limestones.

These three facies interfinger with one another laterally and overstep each other laterally towards the centre of the basin. They will now be described in turn:

Charles facies

These beds are developed around the southern and eastern edge of the Williston basin and extend diachronously basinward over the underlying Mission Canyon facies. Lithologically this facies consists largely of dolomite and anhydrite with minor amounts of halite, shale, and sandstones. These are rhythmically arranged with the following general sequence: each rhythm begins with pelmicrites and biomicrites which grade up into microcrystalline dolomites with scattered shell fragments. These contain anhydrite stringers and nodules towards the top, grading in turn up into anhydrite rocks with dolomite veins.

Mission Canyon facies

The second facies is overlain by, and passes up slope into, the Charles facies, and overlies and passes basinwards into the Lodgepole facies. Petrographically these beds are well-sorted, lime mud-free calcarenites, sometimes dolomitized or sparite-cemented, but often retaining primary interparticle porosity. These rocks are sometimes oolites and sometimes skeletal sands, often composed largely of crinoid debris. Up slope they grade into pelsparites which, with increasing lime mud matrix, pass into the pelmicrites of the Charles facies. The abundant but often fragmented fossils of the Mission Canyon facies include crinoids, brachiopods, bryozoa, corals, foraminifera, and algae.

Lodgepole facies

This facies is best developed in the central part of the basin where it is overlain by the Mission Canyon sediments. Lithologically these rocks are thin-bedded or laminated dark grey argillaceous limestones. They are locally siliceous and interbedded with cherts. The fauna is similar to that of the Mission Canyon facies but much less abundant, better preserved, and with few corals or algae. Fossils are sometimes silicified.

MISSISSIPPIAN (LOWER CARBONIFEROUS) SEDIMENTS OF THE WILLISTON BASIN, NORTH AMERICA: INTERPRETATION

It should be clear from the introductory section of this chapter that the Mississippian deposits of the Williston basin are of carbonate shelf type. The Lodgepole, Mission Canyon, and Charles facies correspond to the X, Y, and Z zones defined by Irwin (1965).

Considered in more detail the fine-grain size and fauna of the Lodgepole facies indicate sedimentation in a low-energy marine environment. The large lateral extent of this facies shows that deposition was in an open sea with sedimentation below wave base, and/or away from bottom currents and terrestrial detritus.

The fragmental fauna, oolites, and clean-washed texture of the

Mission Canyon facies indicate deposition in a high-energy marine environment with winnowing of skeletal sands on migrating shoals and banks. The pelletoidal micrites, crystalline dolomites, and evaporites of the Charles facies suggest deposition in a restricted marine environment cut off from the open sea by bars of skeletal sands. Pelmicrites are typical of Recent lagoons (p. 155). The microcrystalline dolomites and evaporites may have formed by primary precipitation on lagoon floors or by diagenesis in intertidal and supratidal flat deposits analogous to Recent sabkhas as already described (p. 136).

From the stratigraphic relations of the three facies it is apparent that they developed synchronously and migrated basinward one on top of the other. Cyclicity in the evaporite facies may reflect minor fluctuations of the shoreline superimposed on this general basinward regression.

It can be seen therefore that the Mississippian rocks of the Williston basin are a good example of an ancient marine carbonate shelf : shoreline environment.

DISCUSSION OF SHELF DEPOSITS

This section discusses the extent to which the X, Y, and Z zone shelf sea model can be extended beyond the Williston basin, and what modifications, if any, are needed for a wider application.

Carbonates, evaporites, and marls similar to those of the Williston basin occur widely over the northeastern edge of the Arabian Shield and extend into Iraq and Persia. These range in age from Jurassic to Tertiary and are still forming today in the Arabian Gulf.

In the Upper Cretaceous the sea transgressed across the Arabian Shield and into Jordan, Palestine, Egypt, and Libya. Simultaneously marine conditions advanced northwards on the other side of the Tethys geosyncline across Europe. In many regions on both sides of Tethys this transgression deposited a thin glauconite protoquartzite sand blanket (e.g. the Kurnub Sandstone of Jordan and Palestine, and the Greensand of England) which is overlain by predominantly fine-grained limestones. The succeeding formations, which include the Chalk of England, are often composed largely of the microscopic skeletons of planktonic coccoliths, foraminifera, and algal dust. The sparse, but often well-preserved, macrofauna includes lamellibranchs, brachiopods, gastropods, ammonites, echinoids, sponges, corals, and bryozoa. These beds often contain chert layers and, sometimes, phosphates. The stratigraphy of this facies is remarkably uniform laterally, though subtle regional thickness variations can be detected by palaeontological studies. A particular feature of the chalk deposits are 'hardgrounds'. These are scoured surfaces, sometimes bored and phosphatized, overlain by intraformational limestone clasts (Bromley, 1967, p. 507). These provide mute testimony of the uniform history of these deposits. Five 'hardgrounds' can be correlated in a 5-m interval of the chalk for 600 miles across France and England (Jefferies, 1963). Clearly these Upper Cretaceous limestones originated largely below wave base on the floor of an

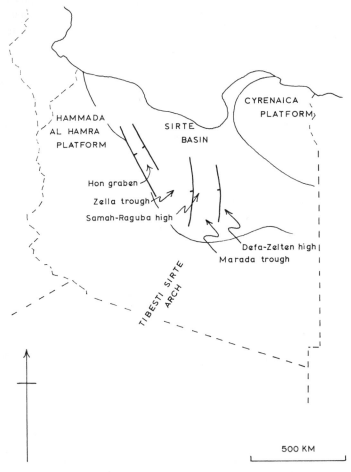

Figure 8.3. Tectonic setting of the Sirte basin, Libya.

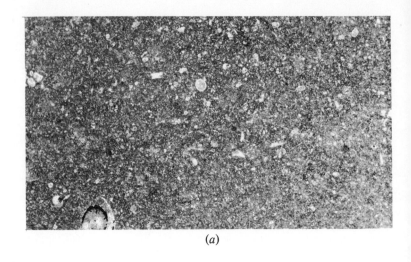

(*a*)

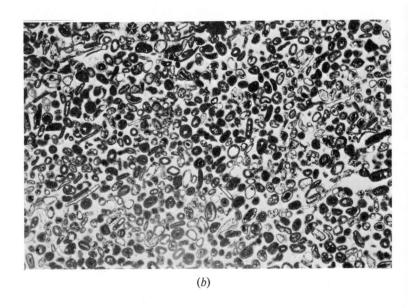

(*b*)

Figure 8.4. Caption on facing page.

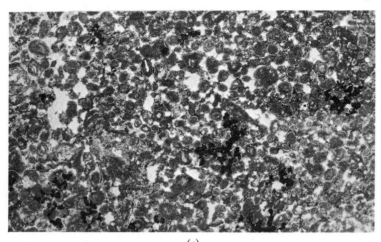

(c)

Figure 8.4. Photomicrographs of Palaeocene carbonate micro-facies from the sub-surface of the Sirte basin, Libya. Released by courtesy of Oasis Oil Company of Libya Inc. (*a*) Calcilutite with pelagic foraminifera, basinal X-zone environment. (*b*) Calcarenite, well-sorted and devoid of micrite matrix, high-energy shelf environment (Y-zone). (*c*) Fenestral limestone of lime mud pellets with sparse lime mud matrix. Uncompacted state of pellets suggests early diagenesis possibly due to penecontemporaneous sub-aerial (?tidal) exposure. Low-energy shelf (Z-zone). Fields of view are 8, 6, and 7 mm wide respectively.

extensive shelf sea with only the gentlest of bottom gradients. They provide a huge example of Irwin's X zone environment.

Towards the end of the Cretaceous Period this sea retreated from Europe but, as already mentioned, persisted over large parts of the Middle East. In some regions, however, this shelf sea was affected by tectonic movements. It is interesting to see how these affected facies patterns. The Sirte basin of Libya is a good case to study. In Cretaceous time the Tibesti–Sirte arch began to fracture and subside into a series of fault-bounded horsts and grabens. Foundering reached its nadir in what is now termed the Sirte basin, or more strictly embayment, since it opened northward into Tethys (Fig. 8.3).

Dark shales and evaporites formed in the grabens and limestones on the horsts. By the end of the Cretaceous this topography was largely subdued though it still controlled the distribution of later sedimentary facies. Palaeocene rocks of the Sirte basin and its environs are of three main types (Fig. 8.4).

Thick laminated grey shales with thin calcilutites were deposited in the centre of the basin and peripheral troughs. These contain a sparse fauna of small pelagic foraminifera such as the Globorotaliids (Fig. 8.4a).

The second facies consists of skeletal calcarenites. These are often devoid of matrix and made of fragments of calcareous algae, bryozoa, corals, large benthonic foraminifera, gastropods, echinoids, and lamellibranchs. There are also several regions where oolites and reefs are present (pages 172, 177). This facies is developed around the rim of the Sirte embayment, notably on the edges of the grabens and on the adjacent shelves (Fig. 8.4b). The third facies extends widely over the shelf areas to the northeast and southwest of the embayment interbedded with skeletal sands similar to those of the previous facies. It consists largely of calcilutites and fenestral limestones. These are pellet deposits which avoided compaction, probably due to early diagenesis, and so retain their original grain textures (Fig. 8.4c). Intergranular sparite cement may be present. Microcrystalline dolomites and evaporites are occasionally found, notably along the northwest flank of the Tibesti–Sirte high.

From this description it can be seen that the three Palaeocene facies of the Sirte basin correspond to the X, Y, and Z zone types of shelf seas defined by Irwin. There are, however, significant differences between the Palaeocene facies and the type examples of the Williston basin. These are largely due to tectonic causes. Unlike the facies belts of the Williston basin those of Libya do not show a simple concentric arrangement with respect to the depocentre. This is due to the palaeohighs which intersect the margin of the Sirte embayment. In such regions facies changes are abrupt, with belts of high-energy Y zone carbonates locally trending directly into the centre of the embayment (e.g. the Zelten–Defa trend). If the palaeohigh was too high, however, the shoal deposits may be absent due to non-deposition, or subsequent erosion. Along the unfaulted edges of the embayment, by contrast, the X, Y, and Z zones are ill-defined laterally with wide transitional zones.

Extensive development of skeletal sands on shelves around the basin margin indicate that the sea bed was within the zone of wave influence. It can be seen, though, that after allowing for tectonic control of facies the X, Y, and Z zone concept is applicable to the Palaeocene rocks of the Sirte basin.

It is possible to extend this concept of high- and low-energy shelf environments outside the realm of carbonates to clastic sediments.

The Lower Palaeozoic rocks of the Saharan and Arabian Shields show a broad sequence of three sedimentary facies (see Bender, 1968; Helal, 1965; Klitzsch, 1966).

(*i*) *Graptolite bearing shales and fine sandstones.* These attain their maximum development in basins, such as the Murzuk and Kufra basins of Libya and the Tabuk basin of Arabia. This facies is generally of Ordovician to Silurian age.

(*ii*) *Medium to fine well-sorted sands* of wide lateral extent. These contain thin shales with *Cruziana* (trilobite tracks), and the burrows *Tigillites* (= *Scolithos* = *Sabellarifex*) and *Harlania* (= *Arthrophycus*). Examples of this facies are the Um Sahm Formation of Jordan and the Haouaz Sandstone of southern Libya (Ordovician).

(*iii*) *Coarse pebbly cross-bedded sands* of wide lateral extent, generally unfossiliferous, e.g. the Saq Sandstone of Saudi Arabia, the Saleb, Ishrin, and Disi Formations of Jordan, and the Hassaouna Sandstone of southern Libya (Cambro-Ordovician).

PreCambrian igneous and metamorphic basement complex.

Sedimentological and trace fossil studies of the above sequences in the Southern Desert of Jordan suggest that the three facies are, from bottom to top, braided alluvial, marine shelf, and deltaic deposits (Selley, 1970; 1972). The supposed marine shelf facies, the Um Sahm Formation of Jordan, will now be described (Fig. 8.5).

Some 250 m thick, this formation crops out widely over the southern desert of Jordan and can be traced eastwards into Saudi Arabia (Helal, 1965). Its southwestward limit is erosional. Northeastwards it is overlain by the graptolitic deltaic shales and sands. The Um Sahm Formation is composed of fine- to medium-grained well-sorted protoquartzites which are dirty white when fresh but weather to a dark brown (Fig. 8.6). The sands are massive or tabular-planar cross-bedded with sets 5–15 cm high. These are grouped in cosets of considerable lateral extent (Fig. 8.7). Unlike the braided alluvial facies beneath (p. 44) they do not occur in channels. Foreset orientations indicate deposition by currents which flowed northeastwards off the Arabian Shield. At several levels there are layers, only a metre or so thick, of laminated grey siltstone and micro-crosslaminated very fine sand. These sequences have sheet geometries

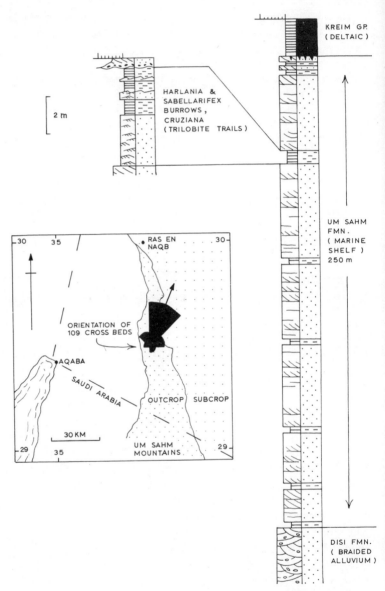

KREIM GP.
(DELTAIC)

HARLANIA &
SABELLARIFEX
BURROWS,
CRUZIANA
(TRILOBITE TRAILS)

2 m

UM SAHM
FMN.
(MARINE
SHELF)
250 m

30   35   RAS EN   30
                           NAQB

ORIENTATION OF
109 CROSS BEDS

AQABA

SAUDI ARABIA

OUTCROP SUBCROP

30 KM

29   29
   35   UM SAHM
             MOUNTAINS

DISI FMN.
(BRAIDED
ALLUVIUM)

Figure 8.5. Summary section and outcrop map of the Um Sahm Formation, Jordan.

and can be traced for kilometres along and perpendicular to the palaeoslope.

Trace fossils in the shales include *Cruziana* tracks, attributed to trilobites, vertical burrows named *Sabellarifex* (= *Scolithos* = *Tigillites*) and sub-horizontal burrows of the type *Harlania* (= *Arthrophycus*).

Figure 8.6. Outcrop of Um Sahm Formation in the Southern Desert, Jordan. Laterally continuous ledges due to soft weathering shale bands with trace fossils.

The cross-bedding and sheet geometry of the Um Sahm sands suggest deposition from megaripples by unidirectional open-flow (i.e. not channel-confined) currents. The shales and rippled sands indicate intermittent low-energy conditions with deposition out of suspension and ripple traction carpets. If the *Cruzianas* are indeed trilobite trails then these waters were probably marine. It seems most probable that the Um Sahm sands were deposited from migrating shoals on an open marine shelf and are therefore a clastic example of Irwin's Y zone environment. The shale units however could be due either to the low-energy offshore sub-wave base X zone, or to the low-energy shallow Z zone to the lee of the sand shoals. The clue to this dilemma

Figure 8.7. Coset of tabular planar cross-bedding in Um Sahm Formation, Jordan.

is provided by the burrows and trails. Studies of Recent and ancient trace fossils show that they can be divided into a number of assemblages specific to different sedimentary environments (Seilacher, 1964; 1967). The assemblage of the Um Sahm shales is a classic example of the near-shore tidal *Cruziana* : *Scolithos* suite. Accordingly the fine-grained sub-facies of the Um Sahm may be attributed to the Z zone environment shorewards of the sand shoals, rather than to a deeper off-shore low-energy X zone.

The Um Sahm Formation of Jordan is therefore a marine shelf deposit. This account demonstrates that the X, Y, Z zone concept can be usefully applied to non-carbonate facies.

To conclude this discussion of marine shelf deposits the following points should be noted.

It is important to distinguish tectonic shelves and basins from sedimentary shelf and basin environments. They are not synonymous.

A tectonic shelf is a stable part of the earth's crust which can be a site of erosion or deposition. Sedimentation may take place in various environments ranging from continental, through shorelines to open marine.

A shelf sedimentary environment occurs on sub-marine tectonic shelves.

A tectonic basin is a large, essentially synclinal, part of the earth's crust (though it may actually be convex to the sky when allowance is made for the curvature of the earth, Dalmus, 1958).

If subsidence occurred faster than sedimentation then deep water deposits may infill the basin (as in the Permian Delaware basin of west Texas, see p. 158).

A slowly subsiding tectonic basin may be infilled by shallow marine shelf, shoreline, or continental deposits.

The open marine shelf environment is therefore only one of several that may occur on a tectonic shelf. It may also prevail over a gently subsiding tectonic basin.

The deposits of a marine shelf are of three main types. Below wave base calcilutites and shales form under low-energy conditions. Where wave base impinges on the gently sloping sea floor high-energy clean-washed sands (carbonate or clastic) form shoals and barriers. Up slope of these a second low-energy environment is found in lagoons and tidal flats. Here deposition may be argillaceous or, if there is no land-derived sediment, pelletal limestones, dolomites, and evaporites may form.

This concept of three main shelf environments is equally applicable to gently subsiding regions such as the Williston basin and to tectonically more complex areas like the Sirte basin. It is relevant to both carbonate environments, as in these two cases, and to clastic shelf deposits such as the Um Sahm Formation of Jordan.

In areas like the Williston basin the evaporites were clearly shallow water (? diagenetic supratidal) deposits which migrated basinwards through time.

Evaporites occur in the centres of many carbonate sedimentary basins such as the Silurian Michigan basin, and the Permian basins of west Texas, and the North Sea. Their situation in such depocentres has led to their being interpreted as originating on the floor of deep restricted marine basins (e.g. Borchert and Muir, 1964, p. 43). It is interesting to speculate, however, whether some of these are actually shallow Z zone deposits of shelf environments (see Shearman and Fuller, 1969).

ECONOMIC ASPECTS

Shelf deposits are of great economic significance. Many of the world's major oil fields occur in shelf carbonates of Iraq, Persia, Libya, and the Williston basin (see Dunnington, 1958; Falcon, 1958; Colley, 1963, and Carlson and Anderson, 1965).

These accounts show that oil accumulates largely in the Y zone shoal carbonates and associated reefs (see p. 176). These often retain their primary porosity (some Libyan reservoirs occur in completely unconsolidated carbonate sand). Alternatively, they may have been cemented but secondary porosity can form by dolomitization and leaching.

The nearshore low-energy Z zone facies generally has less favourable reservoir characteristics. Porosity may sometimes occur in microcrystalline dolomites and pellet limestones. Fracturing can be important in upgrading the reservoir potential of this facies. The caprocks of shelf carbonate reservoirs are generally evaporites, though micritic chalks can also fill this role.

The regionally persistent stratigraphy of shelf carbonates is of great significance since it allows the formation of laterally extensive reservoirs. For example, shoal limestones of the Upper Jurassic Arab-D in Saudi Arabia are host to the Ghawar oil field which is over 100 miles long. The gentle but regionally persistent dips of carbonate shelf deposits allow the local accumulation of hydrocarbon reservoirs from vast catchment areas.

It can be seen, therefore, that carbonate shelf deposits are of some interest in the search for oil and gas. The X–Y–Z concept is a useful exploration tool which aids the prediction of optimum reservoir characteristics in such rocks.

Carbonate shelf deposits are economically interesting for other reasons since they can contain evaporites and phosphates. The origin of phosphates is a complex and controversial problem and the environmental parameters which control their formation are not well understood (e.g. Bromley, 1967). Valuable phosphate deposits occur in Upper Cretaceous shelf carbonates of the Middle East. These are found in a belt stretching from Syria (the Soukhne Formation) through the Wadi Sirhan of Saudi Arabia to Jordan, Palestine, Egypt, and Morocco. Though it is clear from their facies relationships that these phosphates are marine deposits their precise environment of formation seems to be variable. In Egypt, Said attributes them to regressive shoreline conditions (1962, pp. 251, 268) while Youssef attributes them to deposition in syn-sedimentary basins on the deeper parts of the sea bed (Youssef, 1958). In Jordan on the other hand the phosphatic facies is restricted to palaeohighs (Bender, 1968, p. 189). There is an interesting problem here.

REFERENCES

The description of the Williston basin was based on:

Andrichuk, J. M., 1958. Mississippian Madison Stratigraphy and Sedimentation in Wyoming and Southern Montana. In: *Habitat of Oil.* (Ed. L. C. Weeks) Amer. Assoc. Petrol. Geol. pp. 225–67.

Carlson, C. G., and Anderson, S. B., 1965. Sedimentary and tectonic history of North Dakota, part of Williston Basin. *Bull. Amer. Assoc. Petrol. Geol.*, **49**, pp. 1833–46.

Darling, G. B., and Wood, P. W. J., 1958. Habitat of oil in Canadian portion of Williston Basin. In: *Habitat of Oil.* (Ed. L. G. Weeks) Amer. Assoc. Petrol. Geol. pp. 129–48.

Edie, R. W., 1958. Mississippian sedimentation and oil fields in southeastern Saskatchewan. *Bull. Amer. Assoc. Petrol. Geol.*, **42**, pp. 94–126.

Illing, L. V., Wood, G. V., and Fuller, J. G. C. M., 1967. Reservoir rocks and stratigraphic traps in non-reef carbonates. *Proc. Seventh World Petrol. Cong.*, Vol. 2, pp. 487–99.

Irwin, M. L., 1965. General theory of epeiric clear water sedimentation. *Bull. Amer. Assoc. Petrol. Geol.*, **49**, pp. 445–59.

Hansen, A. R., 1966. Reef trends of Mississippian Ratcliffe zone, northeast Montana and northwest North Dakota. *Bull. Amer. Assoc. Petrol. Geol.*, **50**, pp. 2260–8.

Harris, S. H., Land, C. B., and McKeever, J. H., 1966. Relation of Mission Canyon stratigraphy to oil production in north-central North Dakota. *Bull. Amer. Assoc. Petrol. Geol.*, **50**, pp. 2269–76.

Proctor, R. M., and Macauley, G., 1968. Mississippian of Western Canada and Williston Basin. *Bull. Amer. Assoc. Petrol. Geol.*, **52**, pp. 1956–68.

Wendell Smith, G., Summer, G. E., Wallington, D., and Lee, J. L., 1958. Mississippian oil reservoirs in Williston Basin. In: *Habitat of Oil.* (Ed. L. G. Weeks) Amer. Assoc. Petrol. Geol. pp. 149–77.

Other references cited in this chapter were:

Bender, F., 1968. *Der Geologie von Jordanien.* Beitrager Reg. Geol. Erde. Bd. 7, Borntrager, Berlin.

Borchert, H., and Muir, R. O., 1964. *Salt Deposits, the origin, metamorphism and deformation of evaporites.* Van Nostrand, London. 338 p.

Bromley, R. G., 1967. Marine phosphorites as depth indicators. In: Depth indicators in marine sedimentary environments. (Ed. A. Hallam) *Marine Geol.*, Sp. Issue, **5**, No. 5/6, pp. 503–10.

Colley, B. B., 1964. Libya: Petroleum Geology and development. *Sixth World Petrol. Cong. Proc.* Section 1, pp. 1–10.

Dalmus, K. F., 1958. Mechanics of basin evolution and its relation to the habitat of oil in the basin. In: *Habitat of Oil.* (Ed. L. G. Weeks) Amer. Assoc. Petrol. Geol. pp. 883–931.

Dunnington, H. V., 1958. Generation, migration, accumulation, and dissipation of oil in Northern Iraq. In: *Habitat of Oil.* (Ed. L. G. Weeks) Amer. Assoc. Petrol. Geol. pp. 1194–252.

Falcon, N. L., 1958. Position of oil fields of southwest Iran with respect to relevant sedimentary basins. In: *Habitat of Oil.* (Ed. L. G. Weeks) Amer. Assoc. Petrol. Geol. pp. 1279–93.

Helal, A. H., 1965. Stratigraphy of outcropping Palaeozoic rocks around the northern edge of the Arabian Shield (within Saudi Arabia). *Z. Deutsch. Geol. Ges.* Band 117, pp. 506–43.

Holser, W. T., 1966. Diagenetic polyhalite in recent salt from Baja, California. *Am. Mineralogist,* **51**, pp. 99–109.

Illing, L. V., 1954. Bahaman calcareous sands. *Bull. Amer. Assoc. Petrol. Geol.,* **38**, pp. 1–95.

Jefferies, R., 1963. The stratigraphy of the *Actinocamax plenus* subzone (Turonian) in the Anglo-Paris Basin. *Proc. Geol. Ass.,* **74**, pp. 1–34.

Kinsman, D. J., 1969. Modes of formation, sedimentary association and diagnostic features of shallow-water and supratidal evaporites. *Bull. Amer. Assoc. Petrol. Geol.,* **53**, pp. 830–40.

Klitzsch, E., 1966. Comments on the geology of the central parts of southern Libya and northern Chad. In: *South Central Libya and Northern Chad.* (Ed. J. J. Williams) Petrol. Explor. Soc. Libya. pp. 1–19.

Newell, N. D., Purdy, E. G., and Imbrie, J., 1960. Bahaman oolitic sand. *Jour. Geol.,* **68**, pp. 481–97.

Purdy, E. G., 1963. Recent calcium carbonate facies of the Great Bahama Bank. *Jour. Geol.,* **71**, pp. 334–55 and 477–97.

Said, R., 1962. *The Geology of Egypt.* Elsevier, Amsterdam. 377 p.

Seilacher, A., 1964. Biogenic sedimentary structures. In: *Approaches to Paleoecology.* (Ed. J. Imbrie and N. D. Newell) Wiley, N.Y. pp. 296–316.

Seilacher, A., 1967. Bathymetry of trace fossils. In: Depth indicators in marine sedimentary environments. (Ed. A. Hallam) *Marine Geol.*, Sp. Issue, **5**, No. 5/6, pp. 413–28.

Selley, R. C., 1970. Ichnology of Palaeozoic Sandstones in the Southern Desert of Jordan: a study of trace fossils in their sedimentologic context. In: *Trace Fossils.* (Harper, J. C., and Crimes, T. P., Eds.) Lpool. geol. Soc. pp. 477–88.

——, 1972. Diagnosis of marine and non-marine environments from the Cambro-ordovician sandstones of Jordan. *Jl. geol. Soc. Lond.* **128**, pp. 109–17.

Shearman, D. J., 1963. Recent anhydrite, gypsum, dolomite and halite from coastal flats of the Arabian shore of the Persian Gulf. *Proc. Geol. Lond.*, No. 1067, pp. 63–5.

——, 1966. Origin of marine evaporites by diagenesis. *Bull. Inst. Min. Met.*, **76**, section B., pp. 82–6.

——, and Fuller, J. G. C. M., 1969. Phenomena associated with calcitization of anhydrite rocks, Winnepegosis Formation, Middle Devonian of Saskatchewan, Canada. *Proc. Geol. Soc. Lond.*, No. 1658, pp. 235–9.

Youssef, M. I., 1958. Association of phosphates with synclines and its bearing on prospecting for phosphates in Sinai. *Egypt. J. Geol.*, **2**, pp. 75–87.

CHAPTER NINE

REEFS

INTRODUCTION

The term 'reef' was originally applied to rocky prominences on the sea floor on which ships could be wrecked. Coral reefs are a particular form of this navigational hazard found today in tropical waters.

Geologists have applied the term reef to lenses made of the calcareous skeletons of sedentary organisms.

In many cases, however, it is not possible to demonstrate either that an ancient 'reef' was a topographic high on the sea floor or, if it was, that it was wave resistant.

Cummings (1932) classified calcareous skeletal deposits into:

Bioherm 'A reef, bank, or mound; for reeflike, moundlike, or lenslike or otherwise circumscribed structures of strictly organic origin, embedded in rocks of different lithology' (ibid., p. 333).

Biostrome 'Purely bedded structures, such as shell beds, crinoid beds, coral beds, et cetera, consisting of and built mainly by sedentary organisms, and not swelling into moundlike or lenslike forms . . . which means a layer or bed' (ibid., p. 334).

For the purposes of this book the following definitions will be used:

Carbonate build-up: A lens of carbonate rock (a descriptive term). Carbonate build-ups may be divided into two genetic types.

Reef: A carbonate build-up of skeletal organisms which at the time of formation had a rigid framework forming a topographic high on the sea floor.

Bank: A carbonate build-up which was a syn-depositional topographic high of non-wave resistant material, e.g. an oolite shoal, a coquina bank, or a mound of crinoid debris.

This chapter describes ancient carbonate build-ups which have been interpreted as reefs. First, though, the features of Recent reefs will be summarized. Coral reefs are one of the best documented of all present-day environments. The following brief synthesis is based on the references listed at the end of this chapter.

Summary of Recent reefs

The majority of reefs growing at the present time occur in shallow tropical seas. The factors which restrict reef growth are variable but in general they form best in less than 25 fathoms of water where the salinity is between 27–40 per thousand and the temperature seldom drops below about 20°C (Shepard, 1963, p. 351). The inevitable exception to the rule is where coral reefs form in cold water 35 fathoms deep off the Norwegian coast (Teichert, 1958). Reefs of calcareous algae occur in non-marine waters.

The biota of Recent reefs is exceedingly diverse. The resistant reef framework is composed of corals, calcareous algae, hydrocarallines, and bryozoa. Corals are generally actually subordinate to the other groups. Other organisms associated with reefs include calcareous sponges, foraminifera, echinoids, lamellibranchs, gastropods, and sabellariid (carbonate-secreting) worms.

In cross-section a reef complex can be broadly sub-divided into three geomorphological units (Fig. 9.1).

A reef generally shelters a shallow water lagoon from the open sea. The floor of the lagoon is covered by carbonate mud in its deeper parts and sands in shallower turbulent regions. Lagoonal sediments are composed of faecal pellets, foraminiferal sands, coralgal sands of

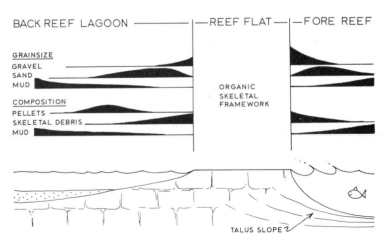

Figure 9.1. Diagrammatic cross-section summarizing geomorphology and sediments of a Recent reef.

comminuted corals and calcareous algae, together with other skeletal sands and finely divided carbonate mud. Scattered through the lagoon may be irregular patch reefs. Grain size increases across the lagoon towards the reef and locally becomes conglomeratic due to organic debris broken off the reef and carried into the lagoon by storms.

The reef itself is, by definition, composed of a resistant framework of calcareous organic skeletons. The top of the reef is flat since the creatures cannot survive prolonged sub-aerial exposure. In addition the upper surface is constantly scoured and planed off by wave action and dissected by seaward trending surge channels. The tops of these sometimes become overgrown to form sub-marine tunnels. The reef framework itself is often highly porous, figures of up to 50 per cent porosity have been cited (Emery, 1956).

The seaward edge of the reef, the reef front, is a sub-marine cliff with a talus slope at its base. This slope is made up of organic debris broken off from the reef front; grain size decreases down slope into deeper water. Immediately at the foot of the reef front boulders of reef rock may be present, grading into sand and then mud in deeper water. The talus slope supports a fauna similar to that of the reef but the quieter conditions allow more delicate branching corals and calcareous algae to form. The reef talus has poorly developed seaward dipping bedding. Slumps and turbidites have been described from ancient reef flanks (p. 164 and p. 187).

On the basis of their geometry recent reefs are classified into three main kinds. Inevitably there are transitions between the three types.

Fringing reefs are linear in plan and stretch parallel to coasts with no intervening lagoons (Fig. 9.2). Fringing reefs can form where low rainfall means that little freshwater and mud is brought into the sea to inhibit the growth of reef colonial organisms. Good examples occur along the desert shores of the Gulf of Akaba on the Red Sea (Friedman, 1968).

Barrier reefs are linear too but a lagoon separates them from the land (Fig. 9.2). This may be narrow or, in the case of the Great Barrier Reef of Australia, an open sea hundreds of miles wide (Maxwell, 1968).

Atolls are sub-circular reefs enclosing a lagoon from the open sea (Fig. 9.2). This kind of reef is abundant in the Pacific; Bikini Atoll is a typical example (Emery, Tracey, and Ladd, 1954). The classic theory for atoll formation was put forward by Darwin in 1842. This proposed that barrier reefs originally formed around volcanic islands. As the island sank under its own weight the barrier would build upwards to keep pace with the relative rise in sea level. Ultimately after all trace of the volcanic island had vanished beneath the sea a circular coral reef, or atoll, would remain. Superimposed on this sequence of events are the effects of Pleistocene sea level changes. Many Recent atolls are developed around the rims of flat-topped plateaux of earlier data.

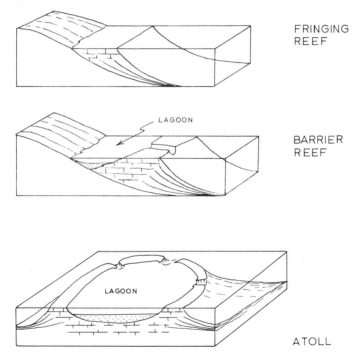

FRINGING REEF

BARRIER REEF

ATOLL

Figure 9.2. Illustrative of the three main types of present day reefs.

Studies of Recent reefs and the concepts derived therefrom can be usefully applied to the study of ancient reefs. Before describing case histories of fossil reefs it is necessary however to make two caution-ary remarks. First, the reef-building organisms of the past were not always of the same groups as those of today. The roles of different groups vary in time and place. For example, calcareous algae have in some situations merely acted as binding agents holding the skeletons of the reef framework organisms together. In other situations and at other times algae formed the dominant framework of reefs.

The second point concerns the relationship between Recent and ancient reef geometries. True fringing reefs appear to be rare in the geological record. Linear reef complexes are common but are generally of the barrier type separating open marine deposits from lagoonal facies.

Fossil atolls are rare. Sub-circular reef complexes have been described as atolls but are not based on volcanic piles. Examples include the Horseshoe Atoll and Scurry-Snyder reefs (Pennsylvanian) of west Texas (Ellison, 1955; Leverson, 1967, pp. 73 and 327) and the El Abra reef of Mexico. This was originally interpreted as a barrier, but offshore studies have now revealed an oval trend (Guzman, 1967).

The second major type of ancient reef, in addition to barriers, is essentially a circular reef core rimmed by reef slope talus. The core may provide shelter for a lagoon on its leeward side. This type of structure is termed a patch reef, or if markedly conical in vertical profile, a pinnacle reef.

Case histories of ancient barrier and atoll reefs will now be described.

PERMIAN REEFS OF WEST TEXAS: DESCRIPTION

One of the best documented ancient barrier reef complexes occurs in Permian rocks of west Texas (Figs. 9.3 and 9.4). These crop out around the rim of the Delaware basin and have been traced eastwards at sub-surface around the edge of the Midland Platform where they contain important oil reservoirs. Because of their economic interest these reefs have been intensively studied at outcrop to aid sub-surface interpretation of facies and trends. The following account is a summary based on the references cited at the end of this chapter. Within the Permian rocks of the Delaware basin area four major sedimentary facies can be defined as follows:

(*i*) Basinal facies.
(*ii*) Shelf margin biolithite facies.
(*iii*) Shelf margin calcarenite facies.
(*iv*) Shelf facies.

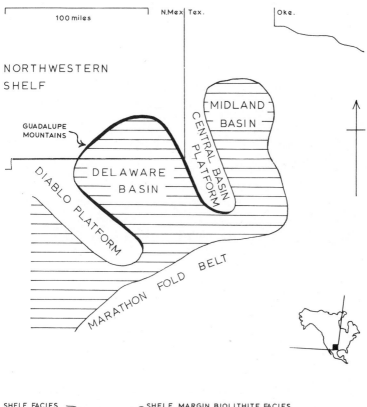

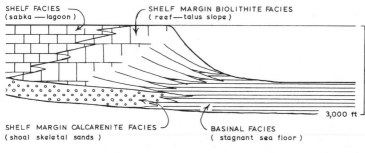

Figure 9.3. Upper: Permian tectonic framework of west Texas. Note areas of reef development (black) around the edge of the Delaware basin. Modified from King, 1934, p. 704, *Bulletin of the Geological Society of America*. Lower: diagrammatic cross-section illustrating facies distribution in the Permian rocks of west Texas. Modified from Newell, N. D., *et al.*, *The Permian Reef Complex of the Guadalupe Mountains, Texas and New Mexico* Fig. 5.D. W. H. Freeman and Company. Copyright 1953.

Figure 9.4. Photograph and sketch of the Guadalupe Mountains, West Texas, showing facies relationships. Photo by courtesy of J. C. Harms, P. N. McDaniel, and L. C. Pray.

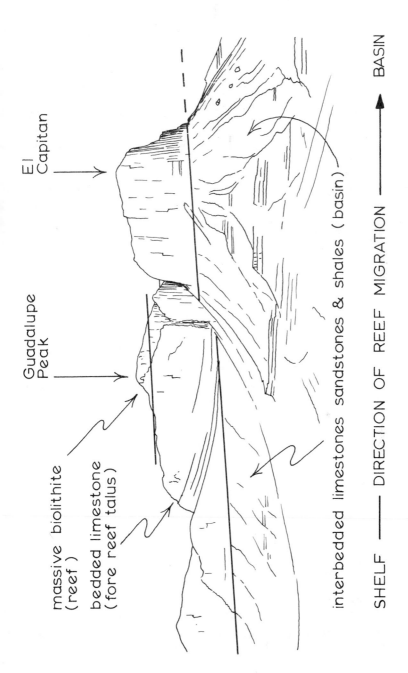

El Capitan

Guadalupe Peak

massive biolithite (reef)

bedded limestone (fore reef talus)

interbedded limestones sandstones & shales (basin)

SHELF —— DIRECTION OF REEF MIGRATION —→ BASIN

The synchronous origin of the facies is shown both by their lateral interfingering and by careful palaeontological control using fusulinid foraminifera. The four facies will now be described in turn.

Basinal facies

Lithostratigraphically this includes the Bell Canyon, Cherry Canyon, and Brushy Canyon Formations of the Delaware Mountain Group. These have an aggregate thickness of several thousand feet and cover some 10,000 sq. miles of the Delaware basin. Lithologically they are composed of thinly interbedded limestones, siltstones, and sandstones.

The sandstones are most abundant in the central part of the Delaware basin. They are grey fine to very fine-grained well-sorted protoquartzites. Individual beds are less than a foot thick with abrupt bases and internal lamination and vertical grading. These features are typical of turbidites (see Chapter 10).

Limestones increase in abundance, bed thickness, and grain size towards the margin of the basin. In the central region they are dark grey or black laminated pyritic siliceous bituminous calcilutites. They contain a sparse benthonic fauna of rhychonelloid brachiopods, nuculoid lamellibranchs, and gastropods such as *Euomphalus*. Pelagic fossils include radiolaria, ammonoids, and the lamellibranch *Posidonia*. Ammonoids often occur crowded on bedding surfaces. Both adult, adolescent, and juvenile forms are found together, suggesting the instant demise of whole tribes of the beasts.

Towards the basin margins the limestones become more abundant, thicker and coarser, grading into packstones (grain-supported calcarenites) and rare calcirudites. Beds are sometimes graded and slumped. These limestones are composed largely of transported fragments of fusulinids, brachiopods, and bryozoa.

Shelf margin biolithite facies

Lithostratigraphically this facies includes the Capitan and Goat Seep formations. These are locally some three hundred feet thick and occur in a belt several miles wide around the edge of the Delaware basin. They can be traced both at outcrop and subcrop for several hundred miles separating shelf and basinal deposits (Fig. 9.3).

Two sub-facies can be defined. A massive limestone overlies and passes basinward into bedded limestone. In both sub-facies primary depositional textures have in many places been destroyed by recrystallization and dolomitization.

The massive limestone consists largely of calcareous fossils, ranging from the *in situ* skeletons of sedentary organisms (biolithite

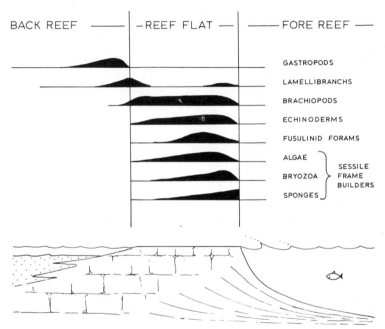

Figure 9.5. Inferred life distribution of west Texas Permian reef organisms. From Newell, N. D., *et al.*, *The Permian Reef Complex of the Guadalupe Mountains, Texas and New Mexico* Fig. 79. W. H. Freeman and Company. Copyright 1953.

or boundstone) to skeletal sands. Interstitial fine grained carbonate is common in all rock-types.

The biolithite is composed of calcareous sponges, bryozoa and hydrocorals bound together in or near growth position by laminae of encrusting calcareous algae (stromatolites). There is a diverse associated fauna which includes fusulinid foraminifera, gastropods, lamellibranchs, brachiopods and echinoderms (largely crinoids).

The biota is zoned ecologically with the sedentary framework builders concentrated on the basinward (seaward) part of the facies (Fig. 9.5).

The massive limestones pass downwards and basinwards into the bedded sub-facies. Bedding planes, dipping as much as 30° into the basin, becoming commoner, gentler and closer spaced down to where they sub-parallel the basinal facies sediments below (Figs. 9.3 and 9.4).

The bedded limestone subfacies contains calcarenites and calcirudites. The calcarenites consist of a poorly sorted framework of skeletal debris and grains of reworked limestone (intraclasts) with abundant interstitial micrite. The fossil fragments are of essentially the same organisms as occur in the massive limestone. Similarly the calcirudites are composed of blocks, sometimes as big as houses, derived from the biolithite. Further evidence of downslope transport of this detritus is provided by slumps and channels.

Shelf margin calcarenite facies

Lithostratigraphically this facies includes parts of the Victoria Peak and Getaway formations, underlying the biolithites separating earlier shelf and basinal sediments. They are laterally discontinuous along the shelf rim.

Lithologically this facies is generally composed of biosparites which are seldom dolomitic or micritic (grainstones). Bedding is well-developed, and cross-bedding is sometimes present. This facies contains fusulinids, branching delicate bryozoa, brachiopods, and crinoids all of which, when comminuted, contribute to the skeletal detritus of the limestones.

Shelf facies

Lithostratigraphically this facies includes the Artesia and Carlsbad Groups. These are about 2,000 ft thick and have sheet geometries which can be traced extensively at outcrop over the northwestern shelf and at sub-surface over the Central Basin Platform (Fig. 9.3).

Any one time interval traced away from the shelf rim shows progressive facies changes from carbonates through evaporites to clastics.

The carbonates adjacent to the biolithite are cross-bedded biosparites sometimes with a derived reef fauna (Newell and others, 1953, p. xv) and elsewhere with a distinctive non-reef fauna (Tyrell, 1969, p. 92). These grade up the shelf into pelletal packstones with thick-shelled molluscs and algae. Locally pisolitic limestones are present. Dolomitization of these beds becomes increasingly common as they are traced up the shelf into laminated algal dolomites with anhydrite nodules (near the surface these are replaced by calcite). The dolomites in turn pass laterally into an evaporitic assemblage of anhydrite, halite, and dolomite. Finally the evaporites interfinger northwards into fine-grained, well-sorted, red rippled, and laminated sandstones and siltstones with desiccation cracks.

As can be seen from Fig. 9.3 the various Permian facies just described migrated through time from shelf to basin. This can be clearly seen from the way in which the massive biolithite overlies the dipping bedded limestones. More dramatic still is the way that the evaporitic and red-bed facies migrate from the back of the shelf out to the centre of the Delaware basin where they constitute the thick Ochoan Series (Hills, 1968, Figs. 4–7). The lower part of the Ochoan Series (the Castile and Salado formations) is composed of several thousand feet of laminated anhydrite, polyhalite, gypsum, halite, and shales. These are often varved, contorted, and with nodular fabrics. Individual shale units can be traced over hundreds of square miles. This evaporite facies is overlain by fine red sandstones and siltstones of the Rustler and Dewey Lake formations.

PERMIAN REEFS OF WEST TEXAS: INTERPRETATION

The biolithite facies of the Guadalupe Mountains contains skeletons of sedentary marine organisms preserved *in situ*, intermingled with fossils of mobile creatures and fine grained detritus. The majority of workers have concluded therefore that this is a fine example of a huge ancient barrier reef in which a reef core of massive limestone prograded seaward over a bedded talus slope, or fore-reef, of its own detritus (e.g. Newell *et. al.* 1953. Silver and Todd 1969, p. 2245).

This interpretation has recently been questioned for two reasons. Achaur, while accepting the *in situ* algal bound nature of the biolithite (1969 p. 2317), claims that this lacked the ecological potential to ever have formed a wave resistant topographic feature (1969 p.

2321). Accordingly he interprets the biolithite facies as a carbonate bank rather than a reef.

Kendall has reached a similar conclusion based on an assessment of biolithite being quantitatively subordinate to detrital limestone in the Capitan Formation (1969 p. 2507).

Because of this uncertainty on the reef/bank origin of the Capitan Limestone it might be thought that this is an inappropriate case history to select as an example of an ancient reef complex. On the contrary it is singularly appropriate since it highlights a current trend to reappraise carbonate build-ups which have previously been designated as reefs. Many have been reinterpreted as banks of non-wave resistant skeletal debris.

Consider now the depositional environments of the facies which are associated with the Capitan Limestone.

The fine grain size and fauna of the basinal facies indicate deposition in a low-energy marine environment. The absence of cross-bedding suggests that this was below wave base and the disseminated pyrite and bitumen indicate reducing anaerobic bottom conditions. Due to the absence of post-depositional tectonics and to careful stratigraphic palaeontology it is possible to calculate the actual depth of the sea floor of the basinal facies. At one point the reef crest was some 1,800 ft above time equivalent basinal sediments (Newell *et. al.*, 1963, p. xvii). The basin floor was therefore well below the depth limit of the present continental shelves (about 600 ft).

The shelf margin calcarenite facies shows by its clean well-sorted texture and cross-bedding that deposition took place from high-energy traction currents. The fauna, albeit fragmentary, indicates a marine environment. Clearly, therefore, this facies was deposited by migrating shoals in the shallow turbulent waters of the shelf edge. The stratigraphic position of these banks show that they predated and formed the foundation of the reefs.

The carbonate shelf deposits immediately to the lee of the barrier reef indicate deposition in a high-energy environment, probably essentially lagoonal. This passed northwards into quieter conditions as shown by the muddy skeletal and pelletoidal limestones. The transition from these beds to laminated dolomites and evaporites which in turn grade into clastics is comparable to sabkha deposits along Recent desert shorelines. The Trucial Coast of the Arabian Gulf is a good analogue (Kinsman, 1964). The pisolitic limestones, originally thought to have formed on the floors of lagoons (Newell *et. al.*, 1953), have recently been interpreted as caliche weathering crusts formed under sub-aerial conditions (Dunham, 1969).

In conclusion the Permian rocks of West Texas are a good example of a carbonate build-up which separated shelf from basinal deposits. On the shelf continental red beds passed seawards into evaporitic sabkhas and algal flats which in turn passed into lagoons. The shelf edge was an area of abundant carbonate sedimentation where both skeletal sand banks and biolithites formed. These prograded seawards over a talus slope of their own detritus onto deep water deposits.

LEDUC (DEVONIAN) REEFS OF CANADA: DESCRIPTION

The Leduc reef complex provides many good examples of patch reefs, the second common type of ancient reefs. These developed on the edges of the Ireton shale basin (Devonian) of Alberta (Fig. 9.6). Patch reefs occur both on the break in slope along the edge of the shelf and on the shelf itself. Since oil was discovered in great quantities in 1947, large amounts of data have been gathered from intensive drilling. The Leduc carbonate lenses overlie the laterally continuous Cooking Lake Formation. This is composed of limestones, sometimes argillaceous, and thin shale interbeds. The limestones are sometimes bioclastic calcarenites and sometimes biostromes with thin bioherms of small lateral extent. The top of the Cooking Lake Formation is a planar surface on which the Leduc Formation carbonate build-ups are based. The crest of individual lenses sometimes lie 1,000 ft above the top of the Cooking Lake Formation (e.g. the Acheson Field area). The majority of the Leduc structures lie along the northeast–southwest slope break on the eastern edge of the Ireton Shale basin. This trend contains the Acheson, Leduc, Bonnie Glen, and Rimby oil fields. The Redwater field occurs in an isolated build-up on the shelf to the east. Another archipelago of reefs of slightly older age occurs on the edge of the Peace River palaeohigh on the northwest slope of the Ireton basin. Petrographically the majority of the Leduc carbonates are medium- to fine-grained vuggy crystalline dolomites and contain few detectable fossils. There are several exceptions to this rule which retain their primary features, one of these will be described in detail shortly. The Leduc carbonate lenses are separated from one another laterally by the Duvernay Formation.

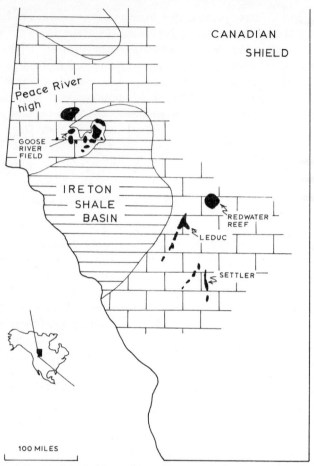

Figure 9.6. Devonian facies distribution in southern Alberta, Canada. Modified from Jenik and Lerbekmo, 1968, Fig. 3. From the *Bulletin of the American Association of Petroleum Geologists*, by courtesy of the American Association of Petroleum Geologists.

This is a thinly-bedded sequence of dark bituminous pyritic limestones and shales. Limestones, largely argillaceous, are best developed in the lower part of the formation and contain a fauna of crinoids, ostracods, and brachiopods. Calcareous shales are dominant towards the top of the Duvernay Formation with a fauna of sponge spicules, brachiopods, ostracods, conodonts, and plant spores.

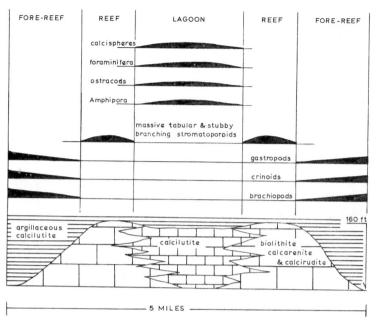

CROSS-SECTION WITH NO VERTICAL EXAGERRATION

Figure 9.7. Cross-section of major sedimentary facies and faunal distribution in the Goose River Reef (Upper Devonian) of Alberta, Canada. Condensed greatly from Jenik and Lerbekmo, 1968, Figs. 8, 9a, and 10a; from the *Bulletin of the American Association of Petroleum Geologists*, by courtesy of the American Association of Petroleum Geologists.

The Leduc and Duvernay Formations are overlain by the Ireton Shale Formation which is locally draped as it thins over Leduc palaeohighs. This formation, over 700 ft thick, consists of laminated pyritic argillaceous dolomites and dolomitic shales with thin shell beds.

The Goose River reef will now be described (Fig. 9.7). This is a good example of the Devonian reefs of Alberta though it slightly predates the Leduc reefs previously described (Fischbuch, 1968). It is atypical of many of these since it is not a crystalline dolomite; primary depositional fabric and fauna have thus been preserved. The following brief account is a summary of an extremely detailed study by Jenik and Lerbekmo (1968).

The Goose River carbonate build-up is one of several that lie on the northwestern shelf of the Ireton basin. It is sub-circular in plan; being about 8 miles in diameter from southwest to northeast, and 5 miles across from southeast to northwest. It is approximately 160 ft high. A total of twenty-two rock types and eight environmental facies have been recognized in this body of rock. For the purpose of this account these have been greatly condensed. Essentially the Goose River build-up consists of a marginal facies, sub-circular in plan, which encloses a central facies.

The marginal facies consists of biolithites, calcarenites, and calcirudites. The biolithite is made up largely of *in situ* stromatoporoid colonies with massive and short stubby branching morphologies. Calcarenites occur as lenses interbedded with the biolithite and in radial channels cross-cutting it. They are composed of fragmented stromatoporoids, bryozoa, brachiopods, and echinoderms (both crinoids and echinoids). Intraformational conglomerates are found on both the outer and inner edges of the marginal facies. These consist of angular and rounded pebbles set in an argillaceous matrix. The pebbles are micritic with scattered entire and fragmented algae, stromatoporoids, bryozoa, crinoids, echinoids, brachiopods, and foraminifera.

The three major lithologies of the marginal facies just described enclose a fine-grained carbonate facies in the centre of the build-up. This consists of laminated micrites, micritic pellet and skeletal sands, and pisolitic and oncolitic algal beds. The biota is different from that of the marginal facies. Macrofossils are largely absent. Microfossils are more common; including foraminifera, ostracods, and calcispheres. (The latter are possibly the reproductive cysts of dasycladacean algae.) Stromatoporoids are present, as in the marginal facies, but have delicate branching morphologies as shown by the form *Amphipora* (Fig. 9.7).

LEDUC (DEVONIAN) REEFS OF CANADA: INTERPRETATION

The geometry of the Leduc Formation shows that it consists of a series of carbonate build-ups, by definition. Where intensive diagenesis has obliterated primary fabrics it is not possible to prove whether these were banks or reefs. It is only in unaltered examples, such as the Goose River build-up, that their true origin can be determined.

The marginal stromatoporoid biolithite facies indicates that the Goose River lens is a true reef. From this initial conclusion the origins of the other facies fall easily into place. The calcirudites and calcarenites associated with the marginal biolithite are probably due to fragmentation of the reef by wave action. The resultant detritus was washed over the reef crest into the lagoon, and allowed to settle on an outer talus slope.

The fine-grained sediments in the centre of the build-up indicate a lower-energy environment in which microfossils and more delicate macrofossils could thrive. The lithology, fauna, and medial distribution of this facies indicate that it formed a lagoon enclosed by the sub-circular reef. Access to the open sea was probably provided by the channels cross-cutting the marginal facies which are infilled by skeletal calcarenites.

A similar, though less symmetrical distribution of facies has also been described from the Redwater build-up on the southeastern shelf of the Ireton basin. In this example a northwest–southeast trending arcuate reef sheltered lower-energy deposits to the southwest. There was no totally enclosed lagoon.

By analogy these two reefs suggest that the other, now recrystallized, build-ups also originated as patch reefs in which arcuate or circular atoll-like reef cores sheltered low-energy environments from the open sea.

GENERAL DISCUSSION OF REEFS

In this section the following three topics are discussed: factors controlling the geometries and facies distribution of reefs, the association of reefs with evaporites and euxinic shales, and lastly, the diagenesis of reef rocks.

Factors controlling reef geometries and facies

The shape of a reef and the distribution of associated facies are controlled by the interplay of sea-level changes, tectonic setting, biota, and oceanography.

It is widely held that ancient reefs grew in shallow water (absolute depth unspecified). This conclusion is based on analogy with Recent reefs. Furthermore, the sedimentary facies associated with ancient reefs generally suggest shallow shelf environments.

If it is true that ancient reefs grew in shallow water their growth must have been closely controlled by fluctuations of sea level. Reef organisms cannot build up above the sea since they are killed by prolonged sub-aerial exposure. Similarly they cannot grow at great depths because algae, being plants, only thrive within the photic zone. Many living ceolenterates are similarly restricted because they have symbiotic relationships with zooxanthellae (a type of algae) within their living tissues.

When the sea level remains static the reef will prograde seaward over its own talus slope, as in the case of the Permian reefs of west Texas. If sea level rises slowly the reef will build essentially upwards with no lateral facies migration, or it will transgress landward over the back reef lagoonal facies. Examples of this occur in the Devonian reefs of the Canning basin, Australia (Playford and Lowry, 1966, p. 8). A rapid rise in sea level will kill a reef due to great depth. A slow drop in sea level will cause the reef to migrate seaward and downward. However, such sequences are relatively rare since, as the shoreline retreats, old reefs tend to be destroyed by sub-aerial erosion. A rapid drop in sea level will kill a reef instantaneously due to prolonged exposure. It can be seen therefore that fluctuating sea level exerts an important effect on the geometry of a reef and its associated facies.

The second main controlling factor of reef geometry is tectonic setting since this is also important in bringing the sea bed in crucial juxtaposition to sea level. Basically reefs are typical of tectonic shelves where sedimentation is shallow, marine, and free from land-derived clastics. Within this broad realm four main sub-types can be recognized. First reefs very commonly form at the edge of a shelf where it passes into a deeper basin. Along such trends barrier reefs may form (e.g. the Permian reefs of west Texas) or discontinuous lines of patch reefs such as the Leduc (Devonian) complex of Canada. Sometimes the edge of the shelf may be a fault with reefs built along the crest of the fault scarp. The Cracoe and Craven reefs of north England may be a case in point (Bond, 1950).

Sometimes a shelf is too deep for reef-formation but local syn-sedimentary movement along anticlinal crests bring the sea bed within a depth sufficiently shallow for reefs to develop. The Palaeocene Intisar reefs of Libya are one example of this (Terry and Williams, 1969, p. 45), the Clitheroe reefs in the Lower Carboniferous Bowland trough of England are another (Parkinson, 1944).

Similarly volcanic eruptions on the sea floor can build piles of lava up into shallower depths where their crests may be colonized by reef organisms. The Devonian Hamar Lagdad reef of the Tafilalet basin, Morocco, is of this type (Massa and others, 1965). As already mentioned, ancient atoll-capped volcanoes are rare. This is possibly because such features are essentially oceanic and unlikely to be preserved in areas at present accessible to the land-based geologist.

The fourth arrangement of reefs on shelves is where patch reefs are distributed at random over a wide area. Such archipelagoes, or buckshot patterns as they are sometimes colourfully called, occur in the Niagaran (Silurian) beds on the edge of the Canadian Shield (Lowenstam, 1950).

The third controlling factor of reef geometry and facies to consider is the biota. The essential control of water depth has been noted already. One of the curious features of reefs is that though they range in age from PreCambrian to the present day, and show similar geometries and sub-facies, their fauna varies through time. PreCambrian algal biostromes have been classified into several types according to their geometry and internal anatomy. Opinions are divided as to whether these differences are of stratigraphic or ecological control (Logan, Rezak, and Ginsburg, 1964). Lower Cambrian rocks in many parts of the world contain reefs built by curious beasts of uncertain affinities termed Archaeocyathines. Reefs are widespread in Palaeozoic rocks throughout the world. The framework of these was built principally by stromatoporoids, rugose corals, hydrocorallines, and bryozoa. Rugose and tabulate corals largely became extinct at the end of the Palaeozoic Era, while the alcyonaria and hexacorals suddenly developed at that point in time and occupied the same ecological niche. In Lower Cretaceous times a particular group of lamellibranchs, the Rudistids, became important reef builders. In these forms one valve became shaped like the calyx of a simple coral while the other formed the lid. Rudistid reefs of Lower Cretaceous age occur in the Edwards Limestone of Texas and the Golden Lane of Mexico (Griffith, Pitcher, and Wesley Rice, 1969), also in the Alps, Libya, and Iran (Henson, 1950). The Richtofenid group of brachiopods followed a similar pseudo-coralline reefal habit in the Upper Palaeozoic. Through the vast span of geological time there seem to have been few aquatic invertebrate groups which have not at some time or another developed colonial reef-forming species. At almost all times, however, the calcareous algae have been important reef organisms (Harlan Johnson, 1961). Generally their role has been to bind together reef frameworks composed of various other organisms. Sometimes, though, they form the framework of reefs largely unaided by other groups. Algal reefs generally tend to be thinner and laterally more continuous than those of other creatures.

Hadding (1950) has described how the geometry of Silurian reefs of Gotland was controlled by their fauna. Three types are present. Lensoid reef knolls are built of spheroidal stromatoporoids, *Favosites* and *Heliolites*. Biostromes are made of tabular stromatoporoids. Pinnacle reefs, with well-developed flank breccias, have cores of delicate branching columnar stromatoporoids.

It can be seen therefore that, since a reef is composed almost entirely of fossils, palaeontology and palaeoecology are vital to an understanding of their depositional environment. Furthermore, since the biota of a reef controls its diagenesis and hence porosity development, an understanding of reef biota is of some economic significance.

The reef : evaporite : euxinic shale association

The Permian rocks of west Texas show a pattern of stagnant basinal sediments rimmed by reefs and overlain by evaporites. This association is a common one in ancient sediments. Weeks (1961) lists nineteen other examples all of which, like the Permian complex, are petroliferous. There has been considerable speculation on the origin and development of such basins. The conventional explanation is that the reefs grew around the margin of a restricted sea where upper waters of normal salinity encouraged reef growth. Dense brines in deeper water inhibited benthonic activity allowing foul-smelling muds to form and ultimately precipitated evaporite minerals on the deeper parts of the basin floor. As the basin filled up the evaporite facies transgressed shorewards killing the reefs and trapping within them hydrocarbons which had migrated from the basinal muds (Hunt, 1967, p. 237). Critical to this concept is the demonstration that marginal reef growth was synchronous with basinal evaporite precipitation (for examples, see Henson, 1950, Fig. 11 and Heybroek, 1965, pp. 28–9).

This is not always possible to prove, however, and an alternative mechanism is sometimes possible. It has already been described how evaporites can form at the present time due to diagenesis of intertidal

and supratidal sabkhas (p. 136). These deposits are not only petrographically similar to those of 'basinal' evaporites but many of the
textures are identical too, such as organic laminations, contortions,
and anhydrite nodules. Thus Shearman and Fuller (1969) have discussed the possibility that inter-reef evaporites of the Devonian
Winnepegosis Formation of Canada may be diagenetic sabkha
deposits rather than basin floor precipitates. If this is correct then the
evaporites formed after the reefs were killed by a drop in sea level.

Similarly, Permian evaporites of the Delaware basin advanced
through time from the shelf to the basin (Hills, 1968, Figs. 4–7).
This is what one would expect from a prograding sabkha. One would
expect the reverse effect if the evaporites formed on the basin floor.
This prompts the question, have tectonic basins (gently subsiding
areas infilled with sediment) been confused with sedimentary basins
(basins which at the time of sedimentation were topographic sub-
marine depressions) (see p. 149).

Clearly the origin of basinal evaporites and their time relationships with reefs deserves careful evaluation.

Diagenesis of reefs

Strictly speaking the diagenesis of reefs is post-depositional and
therefore falls outside the study of depositional environments which
is the theme of this book.

However, reef rock diagenesis is critical to an understanding of
their economic significance. It deserves some attention, therefore.
The diagenesis of carbonates in general, and of reefs in particular, is
a large and complex topic which has been discussed at length in the
literature (see for example Chilingar, Bissell, and Fairbridge, 1967;
Chilingar, Bissell, and Wolf, 1967). The following account is therefore necessarily brief.

Reefs show three interesting and rather unusual properties which
are of great significance to their post-depositional history. First, they
are formed with a very high primary porosity. Second, they are
lithified at their time of formation, compaction is slight and initially
primary porosity is preserved. Three, reefs are formed of chemically
unstable minerals (dominantly aragonite and calcite). These can
undergo diverse chemical changes due to reactions with circulat-
ing pore fluids. Basically two types of diagenetic changes can be
recognized in reefs: mineralogical and textural. These are obviously

inter-related. Initially, aragonitic skeletal material alters to its stable polymorph calcite. Theoretically this should result in an increase in volume and a concomitant decrease in porosity and permeability (Hoskin, 1966, p. 1072). This process takes place at different rates in different areas of the reef complex due to the different stabilities of the various carbonate particles (Friedman, 1964). Simultaneously, or later than the aragonite : calcite change, the carbonates may be enriched with magnesium from the sea water and converted to dolomite. Theoretically this should in contrast result in an overall contraction of the total rock volume of as much as 13 per cent, causing intercrystalline porosity between dolomite crystals (Chilingar and Terry, 1964). Silicification, often involving the selective replacement of fossils, is another diagenetic process common in reefs (Newell *et. al.*, 1953, p. 173).

These chemical changes are associated with textural modification of the reef fabric. First, primary porosity may be reduced shortly after deposition due to the infiltration of finely divided carbonate mud, produced by the fragmentation of calcareous algae. A reduction of porosity may also be achieved through infilling of the framework by sparry calcite accompanied by recrystallization of the carbonate skeletons and the development of calcite overgrowths. These changes all result in a decrease in porosity and, together with dolomitization, often destroy all signs of the original organic fabric. Subsequently secondary porosity may occur due to solution along fractures accompanied by the formation of biomoldic and vuggy porosity. Many of these processes are reversible.

It can be seen therefore that the diagenesis of a reef is complex, being a function of its original faunal and lithological composition and of the chemistry of the fluids which subsequently circulated within it. The main points to be noted, however, are that recrystallization may completely destroy the primary organic fabric of a reef. Secondly, though ancient reefs are often highly porous, the type and distribution of the porosity may bear no relation to that which existed at the time of their formation. These points are critical to an understanding of the economic geology of reefs.

ECONOMIC GEOLOGY OF ANCIENT REEFS

Ancient reef deposits are of great economic significance: because of their porosity they often make good hydrocarbon reservoirs when suitably sealed, and because of their porosity and chemical instability they are liable to replacement by metallic ore minerals.

The association of reefs rimming shale basins and capped by evaporites has been already noted (p. 174). Considerable attention has been given to the occurrence of oil in such situations. It is generally agreed that the evaporites play a critical role both because their parent brines inhibited oxidation of organic matter, allowing petroleum to form, and because the evaporites themselves make good caprocks (Weeks, 1961; Sloss, 1955). The significance of carbonates as oil source rocks in their own right has been frequently discussed (e.g. Hunt, 1967). However, it is generally considered that petroliferous reefs are unlikely to be their own source rocks due to intensive oxidation of organic matter during reef-formation. It is interesting to note, however, that in Tertiary reefs of Iran, oil trapped within fossils predates cementation and is different in composition from that which fills the bulk of the reservoir (Henson, 1950).

Considerable attention has been given to methods of locating oil fields in ancient reefs. This is not always easy. The main problems of identifying ancient reefs are twofold. It is hard to locate reefs since they may be scattered at random over shelves buried beneath horizontal strata often with no surface structural expression. It is hard to identify an oil pool as a reef since diagenesis has often obliterated the original organic fabric. These two problems will now be considered.

Since normal methods of surface geological field work are often inapplicable, reef hunting is largely based on geophysical techniques, particularly seismic shooting. This is often largely a matter of luck. Searching for pinnacle reefs only a mile or two in diameter in several hundred square miles of shelf can be like looking for a needle in a haystack. A case in point was the discovery of the Intisar reefs in the Sirte basin of Libya. The second seismic line to be shot in a 1,880 sq. km concession just happened to transect the crest of a reef. The discovery well was tested for an initial production of 43,000 bd. The seismic programme was initiated to search for structures. The reef was a shock (Terry and Williams, 1969).

Good seismic reflections may delineate reefs when they are evaporite-capped. With other lithologies, however, the reef edge may form a poor reflector, but the reef may be located by a seismic anomaly due to updoming of overlying strata over the crest of the structure (Boulware, 1967, p. 472).

If geophysical data is absent or of poor quality reefs may be located from well data. Criteria to look for are anomalous thickening of limestones between closely spaced wells, anomalous dips due to reef talus bedding, and faunal changes suggesting the proximity of an organic build-up (e.g. Andrichuk, 1958, p. 90). The reef core, if drilled through, may be unrecognizable due to re-crystallization. Where two wells are drilled, one of which passes through an open marine facies and the equivalent interval of the other is lagoonal, it may be profitable to search for a barrier reef in the intervening ground.

Even when a reef has been identified it may be hard to predict the distribution of optimum reservoir properties within it. In some cases production comes from the reef core itself (e.g. the Leduc and Intisar reefs of Canada and Libya). In other situations the reef core may be barren and production restricted to the fore-reef facies as in Tertiary reefs of Iran (Henson, 1950). Back-reef lagoonal facies are generally poor reservoirs. In pelletal limestones porosity may be high but permeability low (Stout, 1964, p. 334). This may be increased by fracturing. Oil production comes from lagoonal sands of the Permian Texas reef complex where they interfinger up dip with evaporites and dolomites (Leverson, 1967, p. 292). Apart from the reef itself, oil and gas may accumulate in overlying non-reef porous strata draped over the reef crest. For additional data on reefal oil reservoirs see Harbaugh (1967, pp. 382–95).

Though oil and gas accumulations in reefs are hard to find and exploit the effort can be rewarding.

Apart from their ability to trap large quantities of oil and gas reefs are also the host to metalliferous deposits. These are of a particular variety termed Mississippi Valley type after one of the prime examples (see Evans, Campbell, and Krouse, 1968, and Brown, 1968). These are telethermal replacement sulphide ore deposits of sphalerite and galena, with subordinate fluorspar, barite, dolomite, and, of course, calcite. Typically these occur replacing reefs and other carbonate build-ups in tectonically stable areas far distant from faults and igneous rocks which could have been sources for the metals. The role of evaporites in providing briny solutions to transport the metals into reefs has been discussed by Davidson (1965). Amstutz in particular has championed a sedimentary origin for the source of many of these ores (Amstutz, 1967, p. 431; Amstutz and others, 1964).

As the mud around a reef compacts, residual metal-rich solutions are expelled and may escape into adjacent porous reefs where they replace the host rock.

As with oil and gas the distribution of minerals within a reef complex varies from place to place. Lower Carboniferous reefs of Ireland have mineralized cores (Derry, Clark, and Gillatt, 1965). The Devonian reefs of the Canning basin, Australia, are mineralized in the fore-reef facies (Johnstone and others, 1967), while in Alpine Triassic reefs mineralization sometimes occurs in the lagoonal back-reef facies.

In conclusion it can be seen that ancient reefs are of considerable economic significance both as hydrocarbon reservoirs and as hosts for mineralization. Sedimentology has a role to play in the location and exploitation of these deposits.

REFERENCES

This is an exceedingly brief bibliography of works on Recent reefs:

Emery, K. O., Tracey, J. I., and Ladd, H. S., 1954. Geology of Bikini and nearby atolls. Pt. I. Geology. *U.S. Geol. Surv. Prof. Pap.*, 260-A, 265 p.

Fairbridge, R. W., 1950. Recent and Pleistocene coral reefs of Australia. *J. Geol.*, **58**, pp. 336–401.

Friedman, G. M., 1968. Geology and geochemistry of reefs, carbonate sediments and waters, Gulf of Akaba (Elat) Red Sea. *J. Sediment. Petrol.*, **38**, pp. 895–919.

Ginsburg, R. N., Lloyd, R. M., Stockman, K. W., and McCallum, J. S., 1963. Shallow water carbonate sediments. In: *The Sea*, vol. III. (Ed. M. N. Hill) Interscience, N.Y. pp. 554–82.

Guilcher, A., 1965. Coral reefs and lagoons of Mayotte Island, Comoro Archipelago, Indian Ocean, and of New Caledonia, Pacific Ocean. In: *Submarine Geology and Geophysics*. (Ed. W. F. Whittard and R. Bradshaw) Butterworths, London. pp. 21–45.

Maxwell, W. G. H., 1968. *Atlas of the Great Barrier Reef.* Elsevier, Amsterdam. 242 p.

McKee, E. D., 1958. Geology of Kapingamarangi Atoll, Caroline Islands. *Bull. Geol. Soc. Amer.*, **69**, pp. 241–77.

——, Chronic, J., and Leopold, E. B., 1959. Sedimentary belts in lagoon of Kapingamarangi Atoll. *Bull. Amer. Assoc. Petrol. Geol.*, **43**, pp. 501–62.

Shepard, F. P., 1963. *Submarine Geology*, Chapter 12. Coral and other organic reefs, pp. 349–70. Harper, London.

Wiens, H. J., 1962. *Atoll environment and ecology*. Yale Univ. Press. 532 p.

The account of the Permian reef complex of west Texas was based on:

Achauer, C. W., 1969. Origin of Capitan Formation, Guadalupe Mountains, New Mexico and Texas. *Bull. Amer. Assoc. Petrol. Geol.,* **53**, pp. 2314–23.

Dunham, R. J., 1969. Vadose pisolite in the Capitan Reef (Permian), New Mexico and Texas. In: Depositional environments in carbonate rocks. (Ed. G. M. Friedman) *Soc. Econ. Min. Pal.,* Sp. Pub., No. 14, pp. 182–91.

Hills, J. M., 1968. Permian basin field area, West Texas and southeastern New Mexico. In: Saline deposits. (Ed. R. D. Mattox) *Geol. Soc. Amer.,* Sp. Pap., No. 88, pp. 17–28.

Kendall, C. G. St. C., 1969. An Environmental Reinterpretation of the Permian Evaporite/Carbonate Shelf Sediments of the Guadalupe Mountains. *Bull. geol. Soc. Amer.* **80**, pp. 2503–2526.

Newell, N. D., Rigby, J. K., Fischer, A. G., Whiteman, A. J., Hickox, J. E., and Bradbury, J. S., 1953. *The Permian Reef complex of the Guadalupe Mountains region, Texas and New Mexico.* Freeman, San Francisco. 236 p.

Rigby, J. K., 1958. Mass movements in Permian rocks of TransPecos, Texas. *J. Sediment. Petrol.,* **28**, pp. 298–315.

Silver, B. A., and Todd, R. G., 1969. Permian Cyclic Strata, Northern Midland and Delaware basins, West Texas and Southeastern New Mexico. *Bull. Amer. Assoc. Petrol. Geol.* **53**, pp. 2223–2251.

Tyrell, W. W., 1969. Criteria useful in interpreting environments of unlike but time-equivalent carbonate units (Tanshill-Capitan-Lamar), Capitan Reef Complex, West Texas and New Mexico. In: Depositional environments of carbonate rocks. (Ed. G. M. Friedman) *Soc. Econ. Min. Pal.,* Sp. Pub., No. 14, pp. 80–97.

The account of the Devonian reefs of Canada was based on:

Andrichuk, J. M., 1958. Stratigraphy and facies analysis of Upper Devonian reefs in Leduc, Settler and Redwater Areas, Alberta. *Bull. Amer. Assoc. Petrol. Geol.,* **42**, pp. 1–93.

Belyea, H. C., 1960. Distribution of some reefs and banks of the Upper Devonian Woodbend and Fairholm Groups in Alberta and British Columbia. *Geol. Surv. Can., Paper, Con., Dept. Mines Tech. Survey,* 59–15. 7 p.

Fischbuch, N. R., 1968. Stratigraphy, Devonian Swan Hills Reef Complexes of Central Alberta. *Bull. Can. Pet. Geol.* **16**, pp. 446–587.

Jenik, A. J., and Lerbekmo, J. F., 1968. Facies and geometry of Swan Hills Reef Member of Beaverhill Lake Formation (Upper Devonian), Goose

River Field, Alberta, Canada. *Bull. Amer. Assoc. Petrol. Geol.*, **52**, pp. 21–56.

Klovan, J. E., 1964. Facies analysis of the Redwater Reef Complex, Alberta, Canada. *Bull. Can. Petrol. Geol.*, **12**, pp. 1–100.

Playford, P. E., and Marathon, O. C., 1969. Devonian carbonate complexes of Alberta and Western Australia: A comparative study. *Geol. Surv. West. Aust.*, Rpt. No. 1, 43 p.

Thomas, G. E., 1962. Grouping of carbonate rocks into textural and porosity units for mapping purposes. In: *Classification of Carbonate Rocks.* (Ed. W. E. Ham) Am. Assoc. Petrol. Geol. Mem. No. 1, pp. 193–223.

Other references mentioned in this chapter were:

Amstutz, G. C., 1967, and Bubinicek, L., 1967. Diagenesis in sedimentary mineral deposits. In: *Diagenesis in Sediments.* (Eds. S. Larsen and G. V. Chilingar) Elsevier, Amsterdam. pp. 417–75.

——, Randohr, P., El. Baz, F., and Park, W. C., 1964. Diagenetic behaviour of sulphides. In: *Sedimentology and Ore Genesis.* (Ed. G. C. Amstutz) Elsevier, Amsterdam. p. 65.

Bond, G., 1950. The Lower Carboniferous reef limestones of northern England. *J. Geol.*, **58**, pp. 430–87.

Boulware, R. A., 1967. Seismic exploration for stratigraphic traps in Canada. *Proc. Seventh World Petrol. Cong.*, **2**, pp. 471–80.

Brown, J. S., 1968, (Ed.). *Genesis of stratiform lead–zinc–barite–fluorite deposits (Mississippi Valley type deposits). Economic Geology – Monograph 3.* Econ. Geol. Pub. Co. Blacksburg, Va. 443 p.

Chilingar, G. V., and Terry, R. D., 1954. Relationship between porosity and chemical composition of carbonate rocks. *Petrol. Eng.*, **B-54**, pp. 341–2.

——, Bissell, H. J., and Fairbridge, R. W., (Eds.). *Carbonate rocks*, Part A. Elsevier, Amsterdam. 471 p.

——, ——, and Wolf, K. H., 1967. Diagenesis of carbonate rocks. In: *Diagenesis in Sediments.* (Eds. G. Larsen and G. V. Chilingar) Elsevier, Amsterdam. pp. 197–322.

Cummings, E. R., 1932. Reefs or Bioherms? *Bull. Geol. Soc. Amer.*, **43**, pp. 331–52.

Darwin, C., 1842. *The Structure and Distribution of Coral Reefs*, London.

Davidson, C. F., 1965. A possible mode of origin of strata bound copper ores. *Econ. Geol.*, **60**, pp. 942–54.

Derry, D. R., Clark, G. R., and Gillatt, N., 1965. The Northgate base-metal deposit at Tynagh, Co. Galway, Ireland. *Econ. Geol.*, **60**, pp. 1218–37.

Ellison, S. P., 1955. Economic applications of Paleoecology. *Econ. Geol.* 50th Anniv., vol. Pt. 2, pp. 867–84.

Emery, K. O., 1956. Sediments and water of Persian Gulf. *Bull. Amer. Assoc. Petrol. Geol.*, **40**, pp. 2354–83.

——, Tracey, J. L., and Ladd, H. S., 1954. Geology of Bikini and nearby atolls. Pt. I. Geology. *U.S. Geol. Surv. Prof. Pap.* **260**-A, 265 p.

Evans, T. L., Campbell, F. A., and Krouse, H. R., 1968. A reconnaissance study of some Western Canada Pb–Zn deposits. *Econ. Geol.*, **63**, p. 349.

Friedman, G. M., 1964. Early diagenesis and lithifaction in carbonate sediments. *J. Sediment. Petrol.*, **34**, pp. 777–812.

——, 1968. Geology and geochemistry of Reefs, Carbonate Sediments and Waters, Gulf of Akaba. *J. Sediment. Petrol.*, **38**, pp. 895–919.

Griffith, L. S., Pitcher, M. G., and Wesley-Rice, G., 1969. Quantitative environmental analysis of a Lower Cretaceous Reef Complex. In: Depositional environments in carbonate rocks. (Ed. G. M. Friedman) *Soc. Econ. Min. Pal.*, Sp. Pub., No. 14, pp. 120–38.

Guzman, E. J., 1967. Reef type stratigraphic traps in Mexico. *Proc. Seventh Petrol. Cong.*, **2**, Elsevier, Amsterdam. pp. 461–70.

Hadding, A., 1950. Silurian reefs of Gotland. *J. Geol.*, **58**, pp. 402–9.

Harbaugh, J. W., Carbonate oil reservoir rocks. In: *Carbonate Rocks*, Part B. (Eds. G. V. Chilingar, H. J. Bissell, and R. W. Fairbridge) Elsevier, Amsterdam. pp. 349–98.

Harlan Johnson, J., 1961. *Limestone-building Algae and Algal Limestones.* Colorado School of Mines. 297 p.

Henson, F. R. S., 1950. Cretaceous and Tertiary reef formations and associated sediments in the Middle East. *Bull. Amer. Assoc. Petrol. Geol.*, **34**, pp. 215–38.

Heybroek, F., 1965. The Red Sea Miocene Evaporite Basin. In: *Salt Basins Around Africa.* Inst. Pet., London. pp. 17–40.

Hills, J. M., 1968. Permian Basin field area, west Texas and Southeastern New Mexico. In: Saline Deposits. (Ed. R. M. Mattox) *Geol. Soc. Amer.*, Sp. Pap., No. 88, pp. 17–28.

Hoskin, C. M., 1966. Coral pinnacle sedimentation, Alacran Reef Lagoon, Mexico. *J. Sediment. Petrol.*, **36**, pp. 1058–74.

Hunt, J. M., 1967. The origin of Petroleum in Carbonate rocks. In: *Carbonate Rocks*, Part B. (Ed. G. V. Chilingar, H. J. Bissell, and R. W. Fairbridge) Elsevier, Amsterdam. pp. 225–51.

Johnstone, M. H., Jones, P. J., Koop, W. J., Roberts, J., Gilbert-Tomlinson, J., Veevers, J. J., and Wells, A. T., 1967. Devonian of Western and Central Australia. In: *Intnat. Symp. Devn. System.* (Ed. D. H. Oswald) Calgary, **1**, pp. 599–612.

Kinsman, D. J., 1969. Modes of formation, sedimentary associations and diagnostic features of shallow-water and supratidal evaporites. *Bull. Amer. Assoc. Petrol.*, **53**, pp. 830–40.

Leverson, A. I., 1967. *Geology of Petroleum*. Freeman, San Francisco. 724 p.

Logan, B. R., Rezak, R., and Ginsburg, R. N., 1964. Classification and environmental significance of algal stromatolites. *J. Geol.*, **72**, pp. 68–83.

Lowenstam, H. A., 1950. Niagaran Reefs of the Great Lakes Area. *J. Geol.* **58**, pp. 430–487.

Massa, D., 1965. Observations sur les series Devonniennes des confins Algero-Marocains du Sud. *C.F.P.Mem.* No. 8, Paris. 187 p.

Parkinson, D., 1944. The origin and structure of the Lower Visean reef knolls of the Clitheroe District, Lancashire. *Quart. J. geol. Soc. London.*, **XCIX**, pp. 155–68.

Playford, P. E., and Lowry, D. C., 1966. Devonian reef complexes of the Canning Basin, Western Australia. *Geol. Surv. West. Aust. Bull.*, No. 118, 150 p.

Shearman, D. J., and Fuller, J. G. C. M., 1969. Phenomena associated with calcitization of anhydrite rocks, Winnepegosis Formation, Middle Devonian of Saskatchewan, Canada. *Proc. Geol. Soc. Lond.* No. 1969, pp. 235–9.

Sloss, L. L., 1959. Relationship of Primary Evaporites to Oil Accumulation. *Proc. Fifth Wld. Petrol. Congr.* Section 1, pp. 123–35.

Stout, J. L., 1954. Pore geometry as related to carbonate stratigraphic traps. *Bull. Amer. Assoc. Petrol. Geol.*, **48**, pp. 329–37.

Teichert, C., 1958. Cold and deep-water coral banks. *Bull. Amer. Assoc. Petrol. Geol.*, **42**, pp. 1064–82.

Terry, C. E., and Williams, J. J., 1969. The Idris 'A' Bioherm and Oilfield, Sirte Basin, Libya – its commercial development, regional Palaeocene geological setting and stratigraphy. In: *The Exploration for Petroleum in Europe and North Africa*. (Ed. P. Hepple) Inst. Petrol., London. pp. 31–48.

Weeks, L. G., 1961. Origin, migration, and occurrence of petroleum. In: *Petroleum Exploration Handbook*. (Ed. G. R. Moody) McGraw Hill, New York. p. 24.

CHAPTER TEN

FLYSCH AND TURBIDITES

INTRODUCTION

This chapter attempts to describe flysch sediments and turbidites. These are particularly interesting rocks which have been written about and argued over for many years. The literature of this problem is vast; whole books have been written on it (e.g. Bouma, 1962; Bouma and Brower, 1964; Dzulinsky and Walton, 1965); accordingly it can only be summarized in this chapter. To begin with, these definitions are important:

Flysch: Thick sequences of interbedded sands and shales. The sandstones generally have erosional bases and are internally graded. The shales contain a marine fauna. First defined in the Alps this word has been applied to similar rocks in geosynclinal belts of all ages and in all parts of the world (for further data see Dzulinsky and Walton, 1965, pp. 1–12).

Turbidity current: 'A current flowing on consequence of the load of sediment it is carrying and which gives it excess density' (Kuenen, 1965, p. 217).

Turbidite: A sediment deposited by a turbidity current.

Greywacke: A poorly sorted sandstone with abundant matrix, feldspar, and/or rock fragments.

For reasons shortly to be described many geologists believe that flysch sandstones are turbidites. Many flysch sandstones, but by no means all, are greywackes. Because of this the terms 'flysch', 'turbidite', and 'greywacke' have been used as synonyms. This is misleading. 'Flysch' describes a facies, 'greywacke' is a petrographic term for a particular kind of rock, and 'turbidite' is a genetic term describing the process which is *thought* to have deposited a sediment. Not only are these three words quite different but they are all extremely loosely defined in the literature.

This chapter begins with an attempt to summarize the characteristics of turbidites. This is followed by a discussion of their origin. The Peira Cava formation, an Alpine flysch sequence, is then described and its environment deduced. The chapter concludes with a summary of flysch and turbidite problems and a discussion of their economic aspects.

DIAGNOSTIC CHARACTERISTICS OF TURBIDITES

Turbidites are identified by no single feature, but by the sum of many criteria. The selection of these is subjective and a function of the prejudices of the individual geologist. An idealized section through a

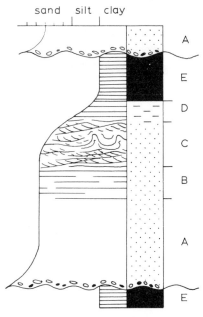

sand   silt   clay

Figure 10.1. Generalized sequence through a turbidite unit. Alphabetic letters code a vertical sequence of five different units (see text). Modified from Bouma, 1962, Fig. 8, by courtesy of Elsevier Publishing Company.

turbidite is shown in Fig. 10.1. Within such a bed it is sometimes possible to detect a wide range of sedimentary structures which are summarized in Table 10.1. This suite of structures has been interpreted in the following way (Walker, 1965; Harms and Fahnestock, 1965; Hubert, 1967). First a current scoured a variety of structures on a mud surface. Sedimentation of the turbidite then took place under waning current conditions. First the massive A unit was deposited, perhaps as antidunes, under an upper flow regime;

shooting flow deposited the next laminated B unit and a lower flow regime deposited the micro-crosslaminated C unit. Various explanations have been offered for the upper laminated silt (Walker, 1965). Perhaps it heralds a return to shooting flow, but since the grain size is much finer, the actual current velocity was probably less than that which caused the lower laminated unit. The upper pelitic E unit indicates resumption of the low-energy environment which prevailed before the turbidite was laid down.

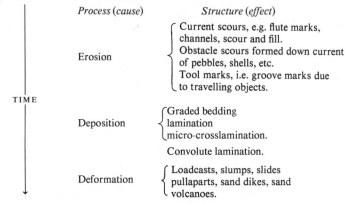

	Process (cause)	Structure (effect)
	Erosion	Current scours, e.g. flute marks, channels, scour and fill. Obstacle scours formed down current of pebbles, shells, etc. Tool marks, i.e. groove marks due to travelling objects.
TIME	Deposition	Graded bedding lamination micro-crosslamination. Convolute lamination.
	Deformation	Loadcasts, slumps, slides pullaparts, sand dikes, sand volcanoes.

Table 10.1. Summary of sedimentary structures associated with turbidites (see Pettijohn and Potter, 1964, for illustrations and definitions).

It is important to note that the above sequence of structures is seldom completely developed in any one turbidite. The previous interpretation is based on the study of many turbidite sequences. Variations on this motif are discussed by Bouma (1962, pp. 49–54), Walker (1965, p. 3), and Van der Lingen (1969, p. 12). Systematic vertical variations in the frequency of bed forms within sequences of turbidites have been documented by Walker (1967). There is a tendency for only the upper units to be developed at the base of a turbidite sequence. Moving up the section, progressively lower units are present within each turbidite until complete A–E sequences appear towards the top of the series. The lower incomplete turbidites, termed 'distal', are believed to have been deposited far from the source. The upper complete units are termed 'proximal' and are believed to have been deposited near the source. A thick sequence of turbidites may thus record a gradual advance of the sediment source into the depositional area.

There has been considerable discussion about the texture of turbidites. Many ancient examples are poorly sorted with clay matrices; petrographically many are greywackes as already mentioned. However, Recent deep-sea sands are widely held to be turbidites, but they are often well-sorted and clay-free (Hubert, 1964). Cummings (1962) has suggested that ancient turbidites might often have been deposited as clean, but mineralogically immature, sands in which the matrix developed from the diagenetic break-down of chemically unstable minerals. Turbidites range in grain size from silts to pebbly sandstones. The coarser varieties, often termed fluxoturbidites, are attributed to deposition by a process midway between turbidity flow and gravity slumping. Petrographically many ancient turbidites are greywackes while Recent deep-sea sands are generally protoquartzites. Protoquartzite ancient turbidites have been identified by Sturt (1961). Carbonate turbidites have been described from ancient and Recent sediments (see Table 10.2). This table also shows that turbidites have been described from a very wide range of depositional

	Recent	*Ancient*
FJORDS:	Holtedahl (1965)	
LAKES:	Grover and Howard (1938)	Kuenen (1951)
DELTAS:	See Chapter 5	See Chapter 5
REEFS:	?	Carozzi and Frost (1966)
CARBONATE SHELF: MARGINS	Rusnak and Nesteroff (1964)	Thomson and Thomasson (1969)
ANCIENT FLYSCH AND RECENT DEEP-SEA SANDS:	See this chapter.	
LAYERED GABBROS:		Irvine (1965)

Table 10.2. Environments from which turbidites have been recorded.

environments and are not even restricted to sedimentary rocks, but have also been described from layered gabbros (Irvine, 1965). It can be seen therefore that turbidites are characterized by the sum total of a wide range of sedimentary structures. Texturally turbidites are generally described from poorly sorted sediments, but this is not always the case. Petrographically they are most commonly recorded from greywackes, but also occur in other sandstone types and in limestones. They have been reported from a wide range of environments. With this great diversity of characteristics one might suppose that turbidites were hard to recognize. A review of the literature suggests that, on the contrary, many geologists think that they can identify a turbidite when they see one with confidence.

DISCUSSION OF THE ORIGIN OF TURBIDITES

It has already been argued (p. 186) that the vertical fining of grain size above a scoured surface suggests that these beds were deposited from waning currents. The suggestion that this current was a turbidity flow was put forward by Kuenen and Migliorini (1950) though the germ of the idea was somewhat older (see Van de Lingen (1969) for a historical review). Subsequently the concept has been greatly elaborated and popularized. The basic idea of the turbidity flow may be summarized thus: In certain regions, such as the edges of continental shelves and deltas, rapid deposition takes place, causing the formation of a steeply sloping pile of waterlogged sediment. Every now and then the slope becomes so unstable that a triggering mechanism, such as an earthquake or storm, starts the sediment moving. Initially it slides and slumps down gullies becoming more and more liquefied until the mixture assumes the characteristics of a turbidity current. That is to say it changes into a muddy liquid whose density is greater than that of the surrounding water. It therefore moves down slope under gravity beneath the clear water. As it gains momentum it becomes erosive, cutting gullies while still on the steep slope and flute marks, etc., on reaching the open-sea floor. On this level surface the velocity starts to diminish. Deposition then takes place, first of the coarsest sediment, and, as the current wanes, of particles of finer and finer grade. When the velocity has returned to zero pelagic mud deposition continues out of suspension as before. All is quiet on the sea floor; a graded bed has been deposited on a scoured surface.

The evidence that turbidity flows deposit the sediments that geologists call turbidites is twofold. It is gathered from experiments and from observations of Recent deep-sea deposits. Graded bedding has been formed from turbidity flows generated artificially in the laboratory (Kuenen, 1948; Kuenen and Migliorini, 1950; Kuenen and Menard, 1952; and Middleton, 1966a, b, and c). Recent deep-sea sands are reported as often showing graded bedding (e.g. Shepard, 1963, p. 406). The suite of internal structures found in ancient turbidites is also sometimes present (e.g. Van Straaten, 1964). The classic oft-quoted reason for believing that Recent deep-sea sands may be turbidites is the case of the 1929 Grand Banks earthquake. This had an epicentre on the edge of the continental shelf off Newfoundland. Telegraphic cables running approximately parallel to the shelf margin were broken in orderly succession downslope. Cores taken subsequently from this area showed a graded bed of silt with shallow water microfossils. This has been interpreted as the product of a turbidity flow which, generated from slumped sediment near the epicentre of the earthquake, ran down the slope ripping up cables and depositing a bed of graded sediment (Heezan and Ewing, 1952; see also Heezan, 1963, p. 743, and Shepard, 1963, p. 339). It has been widely concluded therefore that Recent deep-sea sands are the product of turbidity flows. Ancient sediments which are comparable to deep-sea sands are attributed to the same process and are categorized as turbidites.

This attractive explanation has, none the less, been criticized by a small but articulate group of geologists. Criticism is of two main types, first that deep-sea sands are not turbidites and second that ancient turbidites do not resemble deep-sea sands.

The evidence against a turbidite origin for Recent deep-sea sands has been marshalled by Hubert (1964; 1967), Klein (1967), and Van de Lingen (1969). These authors point out that many deep-sea sands are clean washed, that grading is not always present, and that ripple-marking is common. Traction currents on the ocean floors are strong enough to transport sand, and ripples have been photographed at great depths (e.g. Menard, 1952). It is possible therefore that deep-sea sands are deposited not by catastrophic turbidity flows but by gentle intermittent traction currents. One of the great charms of the turbidite hypothesis was that it could explain the formation of submarine canyons on the edges of the continental shelves. These are now accessible to SCUBA divers and submersibles. Turbidity currents have not been seen in the heads of sub-marine canyons. Instead sand slowly creeps like a glacier, slumping and cascading over hanging valleys. Attempts to generate turbidity flows in canyon heads have been unsuccessful (Dill, 1964).

Evidence against a turbidite origin for ancient rocks which have been called turbidites hinges therefore on the two arguments that they do not resemble Recent deep-sea sands, and that these may be traction deposits anyway. Of particular interest here are the palaeocurrents of ancient turbidites. If these deposits were due to gravity-controlled turbidity flows the orientation of bottom structures and ripple foreset dip directions should coincide and should indicate flow down slope. This is not always the case. Considerable discrepancies have been noted, both between bottom structures and ripple orientations, and palaeoslope, determined from independent criteria (Kelling, 1964, and Klein, 1967). It has been suggested that the reworking of the top of a turbidite by normal oceanic currents may generate ripples migrating in directions oblique to the (?) down slope oriented bottom structures. This is an attractive compromise. However, Hubert (1967) has discovered an Alpine flysch with plant fragment orientations in the shales which coincide with the sand palaeocurrents. Since the former must be due to the normal oceanic currents perhaps the latter are too. Another curious feature is that not all flysch palaeocurrents are slope-controlled as one would expect to find in gravity-controlled turbidites. Scott (1966) found gravity-controlled slumps which had moved at right angles to associated flysch palaeocurrents. In this context it is interesting to note that Recent oceanic bottom currents are not always slope-controlled and often show a tendency to flow parallel to bathymetric contours (Klein, 1967, p. 376).

The arguments for and against a turbidite origin for ancient 'turbidites' and Recent deep-sea sands are endless. From the necessarily short preceding discussion of this problem the following points should be noted: There is a particularly widespread sedimentological phenomenon found in ancient rocks, especially flysch, which is widely interpreted as due to turbidity current deposition. This conclusion is partly based on experimental work and partly on the analogy with Recent deep-sea sands, which are believed to be turbidites. However, nobody has actually observed a turbidite in nature even in the submarine canyon heads which should be their homes. It has been argued that Recent deep-sea sands could be traction-deposited and that, anyway, they are not as similar to ancient turbidites as some authors have suggested. Ancient turbidites sometimes show features that one would not expect to find in a turbidity deposit, notably that they were laid down by currents which were not slope-controlled.

DESCRIPTION OF THE ANNOT TROUGH FLYSCH, ALPES MARITIMES

The case history chosen to document this chapter is a turbidite-rich flysch series from the type area of flysch facies: the Annot trough curves around the southwestern edge of the Argentera–Mercantour massif in the Maritime Alps of southeastern France (Fig. 10.2). It contains over 650 m of flysch of Upper Eocene to Oligocene age. The flysch facies conformably overlies Upper Eocene marls in the trough axis. Laterally equivalent thick sandstone bodies in the trough margins cut into the underlying marls. Two distinct sedimentary facies can be recognized. The axial region of the trough contains the flysch. Around the edge of the trough are tongues of coarser sands and conglomerates (Fig. 10.2). The flysch facies of the Peira Cava area have been studied in great detail. These data will now be summarized. The sequence in this area consists of the classic alternation

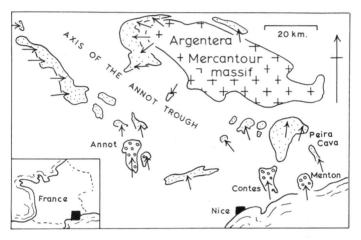

Figure 10.2. Geological sketch map of the Annot trough, Alpes Maritimes. Flysch facies stippled, coarse marginal channel facies shown by circular ornamentation. Arrows indicate palaeocurrents. Modified from Stanley, 1967, Fig. 1.

of sandstones and shales. As the formation is traced northwards into the basin sands become less abundant and their grain size decreases. Within any vertical section the sand : shale ratio increases upward (Fig. 10.3). The sandstones are poorly sorted and range in grain size from coarse to fine. Petrographically they are hard to classify, ranging between arkose and greywacke according to the classification adopted. Sand bed thickness is variable, being generally measurable in decimetres, though beds over 1 m thick are sometimes present. Internally the sands show all the classic features of turbidites. Grading is present in nearly every bed, and the A–E sequence of intrabed units is well developed (Fig. 10.1), being first recognized and defined in this area. The base of the sandstones are erosional with a diverse suite of bottom structures including flute marks, grooves, loadcasts, and scour and fill. Palaeocurrents determined from the sandstones show a generally northerly direction of transport, diverging radially from the southern apex of the present pear-shaped outcrop of the Peira Cava formation.

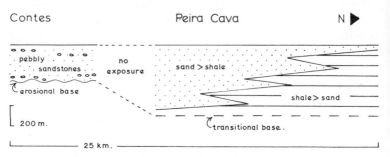

Figure 10.3. Diagrammatic cross-section illustrating facies relationships between the Contes pebbly channel sand and the Peira Cava flysch.

Trace fossils and body fossils are found in the sequence. Both the sands and the shales are sometimes burrowed, and trail covered sand : shale contacts occur. Some rare gastropods, echinoid spines, and faecal pellets have been found. There is an assemblage of pelagic foraminifera.

A second facies is developed in outliers to the south and west of the flysch in the centre of the Annot trough (Fig. 10.2). These outliers occupy north–south trending synclines. The erosional bases and restricted distribution of these patches strongly suggest that their present geometry broadly corresponds to their original distribution. They are believed to have been deposited, therefore, in huge channels up to 5 km wide and, in some cases, with a relief in excess of several hundred metres. Lithologically the sediment filling the channels contrasts markedly with the flysch to the north. Shale is rare, sandstones are coarse, poorly sorted, and pebbly. Bedding is lenticular due to channelling and there is an abundance of slump bedding. Marine shells are present with foraminifera and plant debris. Graded sand : shale units are rare. The orientations of ripples, scour marks, channel trends, and pebble imbrication indicate northerly flowing palaeocurrents.

DISCUSSION OF THE ANNOT TROUGH SANDSTONES

The flysch sandstones of the Annot trough show all the classic features of turbidites discussed at the beginning of this chapter. The coarse marginal channel facies might, at first glance, be thought to be fluviatile. However, the presence of a marine fauna and slumps suggest that these beds were laid down in northerly sloping sub-marine channel complexes. The fact that these merge downcurrent with the flysch facies strongly suggests that the latter were deposited on submarine fans which built out northwards into the deeper parts of the trough. Particularly striking is the relationship between the pebbly sandstone channel complex at Contes from whose northern end the Peira Cava flysch palaeocurrents radiate (Fig. 10.2).

In conclusion this study provides a classic example of a flysch deposited to a large degree by marine slope-controlled turbidity currents. The similarity with Recent sub-marine canyon apron fan complexes is most striking.

FLYSCH AND TURBIDITES: CONCLUDING DISCUSSION

This chapter has described the characteristics of a distinctive sedimentological assemblage of sedimentary structures and graded sandstones which are found in many Recent and ancient sedimentary environments. Such sequences have been widely interpreted as turbidites, due to deposition from muddy sediment suspensions of waning velocity. Evidence for ancient turbidites is based partly on experimental work and partly on observed similarities with Recent deep-sea sands interpreted as turbidites.

Flysch facies are popularly thought to be turbidites. Large volumes of flysch crop out today in mountain chains and are believed to have been deposited in linear troughs termed geosynclines. Many ancient flysch sandstones are poorly sorted with considerable amounts of feldspar and rock fragments. Petrographically such sandstones are greywackes.

There has been considerable discussion of the depth at which turbidites accumulate. As a general rule one might suggest that, though a turbidity flow could occur in shallow water, rapid burial or deposition below wave base would be essential to prevent the turbidite from being re-worked completely by traction currents. As we have seen, however, graded greywackes interpreted as turbidites occur in the Torridonian Series of Scotland interbedded with desiccation-cracked shales (p. 32). Stanley (1968) has reviewed a number of other shallow water turbidites.

The majority of workers, however, have favoured a deep water origin for flysch turbidites. Reasons for this conclusion are largely palaeontological. Many flysch sandstones contain fragments of shallow marine macrofossils, benthonic forams, and plant debris. The interbedded shales however contain pelagic foraminifera which may indicate considerable depths. For example, Natland and Kuenen (1951) described a 19,000-ft turbidite sequence from the Ventura basin, California. At the base foraminifera compare with species living today at depths of 4,000–5,000 ft. Moving up the section foraminifera suggest shallower depths when compared with Recent assemblages. At the top of the sequence the turbidites pass into fluviatile sediments with horse bones. Such studies have been criticized on the grounds that the depth distribution of Recent foraminifera is a function not only of depth but of temperature and other ecological factors (Rech-Frollo, 1962). Two other palaeontological studies bearing on the depth of flysch should be noted. The Jaslo shales (Oligocene) of the Krosno flysch in the Polish Carpathians contain fossil fish with light-bearing organs similar to species living today at oceanic depths (Jerzmanska, 1960). Tertiary flysch from the Pyrenees contains tracks interpreted as bird footprints (Mangin, 1967). Despite this last curious case the weight of the evidence, both faunal and sedimentological, suggests that flysch sands were emplaced at depths below wave base. Deposition was either too deep or too rapid to permit the growth of an abundant benthos. The problem of the depth of deposition of flysch is discussed at greater length by Dzulinski and Walton (1965), Duff, Hallam, and Walton (1967, p. 221), and Van der Lingen (1969).

Criticisms of the turbidite hypothesis put forward to explain flysch have been based on arguments that flysch sands do not resemble deep-sea sands and that, even if they do, the latter may not be turbidites anyway, but due to waning traction currents. Support for this criticism is provided by some palaeocurrent studies which show discrepancies between the palaeoslope and the flysch palaeocurrent indicators. These should coincide if the flysch sandstones were deposited by gravity-controlled turbidity flows.

It is at present difficult to resolve the arguments for and against a turbidite origin of flysch. Despite nearly two decades of intensive research this fascinating facies provides as challenging a problem to the geologist as it has ever done.

ECONOMIC ASPECTS OF FLYSCH AND TURBIDITES

Turbidites deposited in deep marine basins may be interbedded with muds which can be hydrocarbon source rocks. However, turbidite sands do not generally make good reservoirs because poor sorting and clay matrix inhibit porosity and permeability. The thin bedding of turbidite sands and the intervening shales make reservoirs numerous, but thin and disconnected.

Flysch turbidites are very poor oil and gas prospects not just for sedimentological reasons but also because of their restriction to tectonic belts. Here intense folding, faulting, and fracturing allows the escape of all connate fluids.

A case of prolific oil production in turbidites occurs in western California. Here over 30,000 ft of turbidites were deposited in down-faulted Tertiary basins. Subsequent tectonism has involved only gentle folding. As already mentioned (p. 194) foraminifera in the shales indicate original basin depths of some 5,000 ft. These shales are presumably the source rocks. Despite poor sorting and relatively low porosity and permeability the turbidite sands are good oil reservoirs. This is because individual beds are of unusually great thickness for turbidites, sometimes in excess of 10 ft. Furthermore, multistorey sand sequences are common, shale between turbidites being absent either due to erosion or to a rapid frequency of turbidity flows.

Three main geometries can be recognized in Tertiary Californian turbidite facies and these can be related to deep-sea sediments now forming off the present-day Californian coast (Hand and Emery, 1964). Shoestring turbidites occupy channels sometimes located along syn-sedimentary synclines (cf. the Contes channel of the Alpes Maritimes). Sullwold (1961) cites an example 150 ft thick and 2,000 ft wide, whose associated foraminifera suggest deposition in 1,000–2,000 ft of water. The Rosedale Channel (Miocene) of the San Joaquin basin was about one mile wide and contains some 1,200 ft of turbidites. It can be traced downslope for about 5 miles and its fauna suggests depths greater than 1,300 ft (Martin, 1963). Other examples are discussed by Webb (1965).

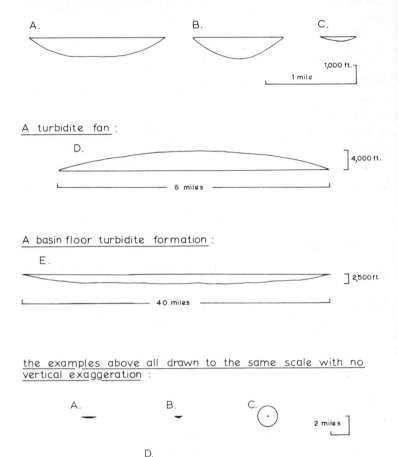

Figure 10.4. Palaeostrike cross-sections of turbidite geometries. Sources of data cited in the text. A. Contes channel, Annot trough, Alpes Maritimes. B. Rosedale channel, San Joaquin basin, California. C. Sansina channel, Los Angeles basin, California. D. Tarzana fan, Los Angeles basin, California. E. Repetto turbidite sheet, Los Angeles basin, California.

Shoestring turbidite bundles such as these probably formed in submarine channels similar to those of the present-day continental margins.

The second turbidite geometry is fan-shaped in plan and lenticular in cross-section. The Upper Miocene Tarzana fan identified by Sullwold (1961) is about 6 miles wide and 4,000 ft thick. Associated foraminifera indicate a depositional depth of about 3,000 ft. Other examples are lobate and branching (Webb, 1965). Geometries such as these can be directly compared to present-day fans formed by sediment debouching from the mouths of sub-marine canyons on to abyssal plains.

The third facies geometry found in Californian Tertiary turbidites is a sheet. This is believed to be due to deposition on a basin floor. The Repetto Formation (Lower Pliocene) is an example of this type. It has a maximum thickness of more than 2,500 ft and covers some 800 square miles (Shelton, 1967).

Sedimentology obviously has a role to play in the exploitation of oil from turbidite facies as, and when, it is found. It is important first to recognize the environment, second to identify the type of geometry, and finally to predict the trend of an individual turbidite bundle.

REFERENCES

The description of the Annot Trough Flysch was based on:

Bouma, A. H., 1959a. Flysch Oligocéne de Peira-Cava (Alpes Maritimes, France). *Eclogae Geol. Helv.*, **51**, pp. 893–900.

——, 1959b. Some data on turbidites from the Alpes Maritimes (France). *Geol. Mijnb.*, **21**, pp. 223–7.

——, 1962. *Sedimentology of Some Flysch Deposits*. Elsevier, Amsterdam. 168 p.

Stanley, D. J., 1962. Etudes sédimentologiques des grès d'Annot et de leurs équivalent latéraux. *Soc. Eds. Technip.* Paris. Ref. 6821, 158 p.

——, 1963. Vertical petrographic variability in Annot sandstones turbidites: some preliminary observations and generalizations. *J. Sediment. Petrol.*, **33**, pp. 783–8.

——, 1965. Heavy minerals and provenance of sands in flysch of central and southern French Alps. *Bull. Amer. Assoc. Petrol. Geol.*, **49**, pp. 22–40.

Stanley, D. J., 1967. Comparing patterns of sedimentation in some ancient and modern submarine canyons. *Preprint Seventh Internat. Sedol. Congress*, 4 p.

———, 1967. Comparing patterns of sedimentation in some modern and ancient submarine canyons. *Earth and Planetry Science Letters*, **3**, pp. 371–80.

———, and Bouma, A. H., 1964. Methodology and palaeogeographic inter-pretation of flysch formations: a summary of studies in the Maritime Alps. In: *Turbidities*. (Eds. A. H. Bouma and A. Brouwer) Elsevier, Amsterdam. pp. 34–64.

Other references cited in this chapter:

Bouma, A. H., and Brower, A., 1964. *Turbidites*. Elsevier, Amsterdam. 264 p.

Cummings, W. A., 1962. The greywacke problem. *Lpool. Manchr. geol. J.*, **3**, pp. 51–72.

Carozzi, A. V., and Frost, S. H., 1966. Turbidites in dolomitized flank beds of Niagaran (Silurian) Reefs, Lapel, Indiana. *J. Sediment. Petrol.*, **36**, pp. 563–75.

Dill, R. F., 1964. Sedimentation and erosion in Scripps submarine canyon head. In: *Marine Geology*. (Ed. R. L. Miller) Macmillan, London. pp. 23–41.

Duff, P. McL. D., Hallam, A., and Walton, E. K., 1967. *Cyclic Sedimentation*. Elsevier, Amsterdam. 280 p.

Dzulinski, S., and Walton, E. K., 1965. *Sedimentary Features of Flysch and Greywackes*. Elsevier, Amsterdam. 274 p.

Grover, N. C., and Howard, C. S., 1938. The passage of turbid water through Lake Mead. *Trans. Amer. Soc. Civ. Eng.*, **103**, pp. 720–90.

Hand, B. M., and Emery, J. O., 1964. Turbidites and topography of North end of San Diego trough, California. *J. Geol.*, **72**, pp. 526–42.

Harms, J. C., and Fahnestock, R. K., 1965. Stratification, bed forms, and flow phenomena (with an example from the Rio Grande). In: Primary sedimentary structures and their hydrodynamic interpreta-tion. (Ed. G. V. Middleton) *Soc. Econ. Min. Pal.*, Sp. Pub., No. 13, pp. 84–115.

Heezen, B. C., 1963. Turbidity Currents. In: *The Sea*, vol. III. (Ed. M. N. Hill) Interscience, N.Y. pp. 742–75.

———, and Ewing, M., 1952. Turbidity currents and submarine slumps and the Grand Banks earthquake. *Am. J. Sc.*, **250**, pp. 849–73.

Holtedahl, H., 1965. Recent turbidites in the Hardangerfjord, Norway. In: *Submarine Geology and Geophysics*. Eds. W. F. Whittard and R. Bradshaw) Butterworths, London. pp. 107–42.

Hubert, J. F., 1964. Textural evidence for deposition for many western North Atlantic deep-sea sands by ocean bottom currents rather than turbidity currents. *J. Geol.*, **72**, pp. 757–85.

——, 1967. Sedimentology of pre-alpine flysch sequences, Switzerland. *J. Sediment. Petrol.*, **37**, pp. 885–907.

Irvine, T. N., 1965. Sedimentary structures in Igneous intrusions with particular reference to Duke Island Ultramafic Complex. In: Primary sedimentary structures and their hydrodynamic interpretation. (Ed. G. V. Middleton) *Soc. Econ. Min. Pal.*, Sp. Pub., No. 13, pp. 220–32.

Jerzmanska, A., 1960. Ichthyo-fauna from the Jaslo shales at Sobniow (Poland). *Acta. Palaeontol. Polon.*, **5**, pp. 367–419.

Kelling, G., 1964. The turbidite concept in Britain. In: *Turbidites*. (Eds. A. H. Bouma and A. Brouwer) Elsevier, Amsterdam. pp. 75–92.

Klein, G. de V., 1967. Paleocurrent analysis in relation to modern sediment dispersal patterns. *Bull. Amer. Assoc. Petrol. Geol.*, **51**, pp. 366–82.

Kuenen, P. H., 1948. Turbidity currents of high density. *Int. Geol. Congr.* 18th Session, pp. 44–52.

——, 1951. Mechanics of varve formation and the action of turbidity currents. *Geol. Fören, Stockholm Förh.*, **73**, pp. 69–84.

——, 1965. Comment. In: Primary sedimentary structures and their hydrodynamic interpretation. (Ed. G. V. Middleton) *Soc. Econ. Min. Pal.*, Sp. Pub., No. 13, pp. 217–18.

——, 1967. Emplacement of flysch-type sand beds. *Sedimentology*, **9**, pp. 203–43.

——, and Migliorini, C. I., 1950. Turbidity currents as a cause of graded bedding. *J. Geol.*, **58**, pp. 91–127.

Mangin, J. P., 1962. Traces de pattes d'oiseaux et flute-casts associés dansun 'facies flysch' dù Tertiare pyrénéen. *Sedimentology*, **1**. pp. 163–166.

Martin, B. D., 1963. Rosedale Channel: Evidence for Late Miocene submarine erosion in Great Valley of California. *Bull. Amer. Assoc. Petrol. Geol.*, **47**, pp. 441–56.

Menard, H. W., 1952. Deep ripple marks in the sea. *J. Sediment. Petrol.*, **33**, pp. 3–9.

Middleton, G. V., 1966a. Small-scale models of turbidity currents and the criterion for auto-suspension. *J. Sediment. Petrol.*, **36**, pp. 202–8.

——, 1966b. Experiments on density and turbidity currents. I. Motion of the head. *Can. Jnl. Earth Sc.*, **3**, pp. 523–46.

——, 1966c. Experiments on density and turbidity currents. II. Uniform flow of density currents. *Can. Jnl. Earth Sc.*, **3**, pp. 627–37.

——, 1967. Experiments on density and turbidity currents. III. *Can. Jnl. Earth Sc.*, **4**, pp. 475–505.

Natland, M. L., and Kuenen, P. H., 1951. Sedimentary history of the Ventura Basin, California, and the action of turbidity currents. *Soc. Econ. Min. Pal.*, Sp. Pub., No. 2, pp. 76–107.

Pettijohn, F. J., and Potter, P. E., 1964. *Atlas and glossary of primary sedimentary structures.* Springer-Verlag, Berlin. 370 p.

Rech-Frollo, M., 1962. Quelques aspects des conditions de dépôt du flysch. *Bull. Geol. Soc. France*, **7**, pp. 41–8.

Rusnak, G. A., and Nesteroff, W. D., 1964. Modern turbidites: terrigenous abyssal plain versus bioclastic basin. In: *Marine Geology.* (Ed. R. L. Miller) Macmillan, New York. pp. 488–503.

Scott, K. M., 1966. Sedimentology and dispersal pattern of a Cretaceous flysch sequence, Patagonian Andes, southern Chile. *Bull. Amer. Assoc. Petrol. Geol.*, **50**, pp. 72–107.

Shelton, J. W., 1967. Stratigraphic models and general criteria for recognition of alluvial, barrier bar and turbidity current sand deposits. *Bull. Amer. Assoc. Petrol. Geol.* **51**, pp. 2441–60.

Stanley, D. J., 1968. Graded bedding–sole markings–graywacke assemblage and related sedimentary structures in some Carboniferous flood deposits, eastern Massachusetts. *Geol. Soc. Amer. Spec. Pap.* No. 106, pp. 211–39.

Sturt, B. A., 1961. Discussion in: Some aspects of sedimentation in orogenic belts. *Proc. Geol. Soc. Land.*, No. 1587, p. 78.

Sullwold, H. H., 1961. Turbidites in oil exploration. In *Geometry of Sandstone Bodies.* (Eds. J. A. Peterson and J. C. Osmond) Amer. Assoc. Petrol. Geol. pp. 63–81.

Thomson, A. F., and Thomasson, M. R., 1969. Shallow to deep water facies development in the Dimple limestone (Lower Pennsylvanian), Marathon Region, Texas. In: Depositional environments in carbonate rocks. (Ed. G. M. Friedman) *Soc. Econ. Min. Pal.*, Sp. Pub., No. 14, pp. 57–78.

Walker, R. G., 1965. The origin and significance of the internal sedimentary structures of turbidites. *Proc. Yorks. Geol. Soc.*, **35**, pp. 1–32.

——, 1967. Turbidite sedimentary structures and their relationship to proximal and distal depositional environments. *J. Sediment. Petrol.*, **37**, pp. 25–43.

Webb, F. W., 1965. The stratigraphy and sedimentary petrology of Miocene turbidites in San Joaquin Valley (Abs.), *Bull. Amer. Assoc. Petrol. Geol.*, **49**, p. 362.

Van der Lingen, G. J., 1969. The turbidite problem. *N.Z. Jl. Geol. Geophys.*, **12**, pp. 7–50.

Van Straaten, L. M. J. U., 1964. Turbidite sediments in the southeastern Adriatic sea. In: *Turbidites.* (Eds. A. H. Bouma and A. Brouwer) Elsevier, Amsterdam. pp. 142–7.

CHAPTER ELEVEN

PELAGIC DEPOSITS

INTRODUCTION: RECENT PELAGIC DEPOSITS

The previous chapters of this book have been concerned with the diagnosis of sedimentary environments which were either continental or, if marine, of continental shelf origin. Turbidites were an exception to this, their depth being diverse, but, in the case of flysch deposits, possibly below the continental shelf.

This last chapter is concerned with ancient sediments which have been attributed to a pelagic environment. This 'is generally applied to marine sediments in which the fraction derived from the continents indicates deposition from a dilute mineral suspension distributed throughout deep-ocean water' (Arrhenius, 1963, p. 655). To prove that an ancient sediment was deposited in a pelagic environment, as defined above, it is necessary to demonstrate both an absence of terrestrial influence and great depth. As this chapter shows it is not easy to prove both these points in ancient sediments. It may be safer therefore to define pelagic deposits as 'of the open sea' (Riedel, 1963, p. 866).

The need to make this distinction between depth and 'oceanicity' in analysing the environment of supposed ancient pelagic deposits has been stressed by Hallam (1967, p. 330). The Recent seas can be divided into those of the continental shelves which are separated by the shelf break, at about 600 ft from the oceans.

Deposits of the Recent oceans are currently being studied intensively (see Shepard, 1963, Chapter 14, and Arrhenius, 1963, for reviews).

Recent oceanic sediments can be broadly classified into these types:

(*i*) Terrigenous sediments.
(*ii*) Calcareous oozes.
(*iii*) Siliceous oozes.
(*iv*) Red clays.
(*v*) Manganiferous deposits.

These will now be briefly described and their distribution discussed.

Terrigenous sediments are present adjacent to the continents. They consist of argillaceous deposits and the deep-sea sands, of possible turbidite origin, discussed in the previous chapter. Calcareous oozes, or muds, are composed largely of the shells of microfossils. Two main types may be distinguished: Pteropod ooze made up largely of the aragonitic shells of this mollusc, and foraminiferal oozes, composed largely of calcitic foraminiferal tests, often of *Globigerina*.

Siliceous oozes are made of the skeletons of Diatoms and Radiolaria.

Red clays are red and dark brown muds which are believed to form from the finest of wind-blown dust from continental deserts, together with extremely fine particles of volcanic ash and cosmic detritus.

The last type of pelagic deposit is diagenetic rather than depositional in origin, being scoured surfaces impregnated and overlain by nodules rich in manganese. It has been estimated that this is the most extensive type of hard rock surface on the lithosphere (Mero, in Shepard, 1963, p. 404).

The distribution of the various Recent pelagic deposits are roughly correlated with depth. Thus over much of the Atlantic and Pacific oceans Red clays occur in the deepest parts, Radiolarian oozes form in shallower waters deeper than about 4,500 m, calcitic foraminiferal oozes lie on the ocean bed between 4,500 m and 3,500 m, above which point aragonitic pteropod and foraminiferal oozes are found. Though the distribution of pelagic deposits is loosely correlated with depth it is actually controlled by a variety of factors such as rate of sedimentation and rate of solution. Thus the sequence, with increasing depth, of aragonitic, calcitic, siliceous, and argillaceous deposits largely reflects their increasing chemical stability. The rate of solution of these minerals is a function of their rate of burial, the water temperature and its state of saturation by the various chemicals, as well as the hydrostatic pressure. Only the last of these is actually depth-dependent. This point is illustrated by the fact that, regardless of depth, calcareous oozes are poorly developed in polar regions. This is due to the low temperature of the bottom water which causes the solution of aragonite and calcite at faster rates than in the more equable equatorial oceans.

This fact, that the distribution of Recent pelagic deposits is not a direct reflection of depth, is critical to attempts to determine absolute depths of ancient pelagic deposits.

At the present time the oceans cover much more of the earth's surface than the continental regions. There does not appear to be a corresponding dominance of deep over shallow water sediments in the geological column; on the contrary, ancient deep-sea deposits appear to be very rare. This supports the concept of isostasy since it suggests that the continents have never been submerged to great depths below the oceans. Not only are ancient deep-sea deposits very rare but their depth is often debatable. This is because it is very hard to separate criteria of depth of deposition from those which indicate distance from the land. The two are not synonymous as the wide continental shelves of the Atlantic indicate today. This dilemma will become apparent in the discussion of the case history which follows.

PELAGIC DEPOSITS OF THE ALPS: DESCRIPTION

Lengthy geological studies have revealed that during Mesozoic and early Tertiary time marine sediments were laid down in a trough stretching from southern France, northern Italy, and beyond to the east. In mid-Tertiary time these deposits were folded and uplifted to form the Alps and associated mountain chains. This trough is called the Tethys geosyncline. In general the following sequence of events seems to have taken place in this region:

(*i*) Deposition of marine carbonates and fine-grained clastics (Triassic–Jurassic); thought to be shelf sediments.

(*ii*) Deposition of thin sequences of clays, cherts, and limestones, sometimes associated with vulcanicity (Triassic–Lower Cretaceous); thought to be pelagic in origin.

(*iii*) Deposition of thick flysch sand : shale sequences (Cretaceous–Oligocene); thought to be turbidites.

(*iv*) Phase of folding, thrusting and uplift.

(*v*) Deposition of clastic sediments in troughs marginal to the mountains, e.g. the Swiss plain and the north Italian plain. This, the molasse facies, is generally continental, at its base is locally marine and sometimes calcareous (Oligocene–Recent).

The timing of these events varies considerably from place to place; two phases often occurring simultaneously in adjacent areas. The clays cherts and limestones of the second phase are often cited as examples of deep-water deposits. They will now be described and the reasons for this interpretation will be discussed. These deposits may be divided into three sub-facies defined as follows:

(*i*) Clay shales, marls, and micritic limestones.
(*ii*) Radiolarian cherts.
(*iii*) Nodular red limestones.

Clay shales, marls, and micritic limestones

This sub-facies consists of clay shales, sometimes dark and pyritic, sometimes calcareous, marls, and fine-grained micro-crystalline limestones. These beds are generally massive or laminated. The various lithologies sometimes occur rhythmically interbedded with one another. Fossils are generally rare but often well-preserved and sometimes crowded on bedding surfaces. They include thin-shelled lamellibranchs termed *Posidonia* or *Bositra*, belemnites, ammonites, and brachiopods. Often ammonite shells are absent but the lids of their shells (aptychi) occur. A particularly characteristic brachiopod of this facies is *Pygope*, an unusual form with a hole in the middle. Beds of this sub-facies are named according to their fauna, e.g. Aptychus marls, Pygope limestone, and so on. It is widely distributed in Upper Jurassic and Lower Cretaceous rocks of the Alps, the Apennines, and Greece.

Radiolarian cherts

Radiolarian cherts are often associated with the above sub-facies and share a similar geographic and stratigraphic distribution. They are generally thin-bedded and dark grey, black, or red in colour. Bedding is often rhythmic with thin shale partings. Siliceous limestones are sometimes present transitional between the cherts and the micrites of the previous sub-facies. The cherts are often largely composed of the microscopic siliceous tests of Radiolaria. In addition they contain minor quantities of silicified sponge spicules and foraminifera. The macrofauna is similar to that of the previous sub-facies but is generally much rarer. This chert facies has been considered as an insoluble residue from which all calcium carbonate has been dissolved.

Nodular red limestones

Red nodular limestones, termed 'ammonitico rosso', occur associated with the two previous sub-facies and are widespread over the Alps,·

Figure 11.1. Polished slab of Ammonitico Rosso, Adnet Limestone, Lower Lias, Adnet, Austria. Pale pink micritic limestone fragments set in a matrix of red clay rich in iron and manganese. From Hallam (1967 Plate 2. Fig. 1.) by courtesy of the Editors of the Scottish Journal of Geology.

Apennines, and parts of Greece. They range in age from Triassic to Jurassic. Lithologically this sub-facies consists of beds of pink micrite nodules tightly packed in a dark red marl matrix. The nodules are often coated with iron and manganese (Fig. 11.1). Thin beds of manganese nodules are sometimes present. The characteristic fossils of this facies are the ammonites from which it gets its name. These are locally abundant. Also present are the aptychi of ammonites, belemnites, crinoids, echinoids, lamellibranchs, gastropods, foraminifera, and sponge spicules. These fossils often show signs of corrosion. Scoured bedding surfaces, termed 'hard grounds', are sometimes present. These are generally very rich in iron and manganese and are occasionally phosphatized. These surfaces are sometimes pierced by *Chondrites* burrows and overlain by a thin layer of manganese or phosphate nodules.

PELAGIC DEPOSITS OF THE ALPS:
DISCUSSION OF ORIGIN

The three sub-facies just described have a number of points in common. They all show an absence of land-derived quartzose sediment and have obviously marine faunas. Due to complex tectonics, their stratigraphic relations are often uncertain, but palaeontology indicates that they were deposited at the same time as adjacent thicker shelf sediments whose depth must be relatively shallow since they are sometimes reefal. These rocks are often overlain by flysch turbidites. All the sub-facies are fine-grained and devoid of sedimentary structures such as cross-bedding and channelling which would suggest strong current action. The 'hard grounds' may be due to gentle scouring of early cemented sea beds. The fauna is characterized by an abundance of pelagic fossils and a scarcity of benthos. Though the shells are generally preserved as calcite, as are most fossils, there is a curious absence of shells of original aragonitic composition. Thus the calcitic aptychi of ammonites often occur while their aragonitic shells do not. Chemically the rocks are unusual due to the abundance of organic silica and calcite, together with the presence of unusual amounts of iron, phosphate, and manganese.

The fine grain size and absence of cross-bedding and channelling indicate that these are low-energy deposits which must have formed below wave base, though as the 'hard grounds' show, subjected to intermittent scouring by currents.

According to the deductions just made these deposits are of pelagic origin, in the sense that they originated in the open sea far from continental influence. There is, though, some controversy as to the precise depth of their deposition. The evidence cited in favour of great depth hinges on the analogy of these sediments with Recent deep-sea deposits. Similarities are to be found in the absence of land-derived clastics, the scarcity of current structures, and scarcity of benthos, despite the frequent oxygenation of the sea bed, indicated by the presence of red ferric iron. The limestones invite comparison with Recent Globigerina oozes, the cherts with Recent radiolarian oozes, and the manganiferous beds with the red clays and manganiferous nodules. The lack of fossils of aragonitic shells can be attributed to the greater solubility of aragonite over calcite. At the present day aragonite goes into solution at depths of about 3,500 m (Friedman, 1965) and calcite at about 4,500 m (Berner, 1965). Hard ground horizons occur between 200 and 3,500 m (Fischer and Garrison, 1967).

All these analogies suggest a deep water origin for the Alpine beds described above. On the other hand it can be argued that many of these criteria are either dubious on chemical grounds, or that they indicate distance from the land, not great depth. This is particularly valid in the case of the absence of land-derived quartz and the abundance of fine-grained clay, and organic calcareous and siliceous sediment. Geochemical studies of Recent deep-sea nodules have laid stress on their oxygen, iron, and manganese ratios and on their trace element content. Recent manganese nodules occur today in widely varying depths and even in some lakes. Their chemical variation may be not just depth-controlled but also due to proximity of land-derived chemicals (Price, 1967). Similarly the solution rates of aragonite and calcite are controlled not just by depth, but also by temperature, as pointed out in the introduction to this chapter. Sedimentation rate will also control solution, slow burial increasing the time available for a shell to corrode on the sea bed (Hudson, 1967).

The facies relationships of the supposed deep-water deposits are sometimes curious too. For example in the Unken syncline of Austria there is a 300-m sequence of Jurassic pelagic limestones, radiolarian cherts, and ammonitico rosso for which Garrison and Fischer (1969) have discussed depths in the order of up to 4,000–5,000 m. This sequence overlies a karstic topography of Upper Triassic limestones and at the top is unconformably overlain by Lower Cretaceous (Berriasian) limestones. There must have been quite a tectonic flutter between the end of the Triassic and the Lower Cretaceous.

In conclusion the fine-grained limestones, ammonitico rosso, and radiolarian cherts of the Alpine geosyncline are clearly marine deposits. They were laid down in a low-energy environment below the level of strong current activity and, as the absence of algae show, below the photic zone. The condensed stratigraphy shows that sedimentation was slow, with little influx of land-derived material. They may be attributed to a pelagic environment with confidence. Comparison with Recent deep sea deposits suggests a deep-water origin. However the distribution of these is not directly depth-dependent, many of the similarities may be controlled, not necessarily by depth, but by distance from the land. This comparison may not therefore be valid.

GENERAL DISCUSSION OF PELAGIC DEPOSITS

Whether the deposits described above are deep water or not one very important point must be noted. Pelagic facies are often present in a similar position in other ancient geosynclinal sequences. This can be seen in Upper Palaeozoic sediments of the Variscan geosyncline which extended from southern Ireland and southwestern England eastwards through Germany. Devonian and Carboniferous rocks of these regions often consist of laminated black shales with ostracods and trilobites, radiolarian cherts, and red nodular cephalopod-bearing limestones not unlike the ammonitico rosso. Goldring (1962) has produced an elegant study combining absolute age determinations with the thicknesses of different facies in various regions. This enables an estimation of sedimentation rates to be made. The data suggest that deposition slowed markedly when the Old Red Sandstone fluviatile sedimentation was followed by the 'bathyal' pelagic facies in the axis of the geosyncline. Sedimentation speeded up again at different times in different regions when the bathyal phase was succeeded by flysch of the Culm Series.

A further common characteristic of ancient early geosynclinal pelagic sediments is their association with volcanic activity. This typically consists of diverse pillow lavas, spilites, basalts, and serpentinites. This association, generally termed the Ophiolitic Suite, is found in the Jurassic and Cretaceous rocks of Greece, the Apennines, and the Alps. It occurs in Upper Palaeozoic rocks of the Variscan geosyncline in southwest England and Germany, and is present in the Lower Palaeozoic Caledonian geosyncline, notably in the southern uplands of Scotland. There has been considerable speculation that sub-marine volcanic eruptions locally increased the silica content of sea water, encouraging population explosions of radiolaria. Aubouin (1965) has shown, however, that radiolarian chert formation was independent of vulcanicity in the Jurassic and Lower Cretaceous sediments of the Pindus furrow in Greece. It might be interesting to relate the present distribution of radiolarian oozes to sub-marine vulcanism.

Finally one more example of possible ancient deep-water sediments deserves mention. This occurs in the island of Portuguese Timor between Australia and southeast Asia. Here some curious beds of Upper Cretaceous age lie beneath a thick Tertiary flysch sequence. In western Timor these contain red clays with manganese nodules, sharks teeth, and rare fish bones. In eastern Timor there is a diverse assemblage of radiolarian cherts, calcilutites, marls, turbidites, and shales rich in iron and manganese and containing manganese nodules. These sediments have been the subject of a detailed study by Audley-Charles (1965; 1966). The nodules of western Timor are geochemically similar to Recent deep-sea nodules while the eastern Timor examples show characteristics midway between Recent deep-sea and shelf nodules. Here again, however, the direct correlation of depth and geochemistry has been questioned by Price (1967) who, as previously mentioned (p. 207), has suggested that the chemical variation of modern manganese nodules is controlled by oceanicity rather than depth.

To conclude this discussion of ancient pelagic sediments the following points seem to be important. Turbidite flysch sequences often overlie a distinctive facies which can be recognized in various parts of the world in rocks of varying ages. This is composed of fine-grained limestones, shales, and marls. The limestones are sometimes red and nodular. They are associated with radiolarian cherts and, sometimes, with sub-marine vulcanicity. The fauna of this facies is typically pelagic; benthonic fossils are generally extremely rare. Sedimentary structures suggesting strong current action are lacking. Sequences of this facies generally show few stratigraphic breaks and are thinner than nearby time-equivalent facies.

It is clear that such rocks were slowly deposited in marine environments below the photic zone and away from strong current action. In many ways these sediments are comparable to Recent deep-sea sands. Is this similarity due to the fact that they both formed in deep water, or due to distance from the land?

ECONOMIC SIGNIFICANCE OF DEEP-SEA DEPOSITS

Fortunately, since their origin is often debatable, ancient deep-sea deposits are of relatively little economic significance. They are generally of low primary porosity though secondary fracture porosity may be present especially in the cherts. Since deep-sea deposits are often oxidized they are unlikely to be hydrocarbon source rocks. The restriction of deep-sea sediments to the axes of mountain belts renders them improbable as hydrocarbon sources or reservoir rocks, since tectonic activity is likely to have permitted the escape of all indigenous fluids and gases.

However, the volcanic rocks associated with deep-sea sediments may be metalliferous and sub-marine eruptions can cause the local precipitation of economically valuable minerals on the sea bed. The metalliferous brines in the 'hot holes' of the Recent Red Sea may be a present day analogue of this kind of situation (Degens and Ross, 1969).

An example of an economic mineral concentration in ?deep-sea sediments are certain barytes deposits of Nevada. These occur in an Ordovician geosynclinal sequence of radiolarian cherts and shales. Barytes is interbedded with these and forms thin conglomerates mixed with phosphate nodules. The presence of re-worked barytes fragments strongly supports its primary origin (Shawe, Poole, and Brobst, 1969).

REFERENCES

The description of Alpine deep-sea deposits was based on:

Aubouin, J., 1965. *Geosynclines*. Elsevier, Amsterdam. 335 p.

Garrison, R. E., and Fischer, A. G., 1969. Deep water limestones and radiolarites of the Alpine Jurassic. In: Depositional environments in carbonate rocks. (Ed. G. M. Friedman) *Soc. Econ. Min. Pal.*, Sp. Pub., No. 14, pp. 20–56.

Hallam, A., 1967. Sedimentology and palaeogeographic significance of certain red limestones and associated beds in the Lias of the Alpine regions. *Scott. J. Geol.*, **3**, pp. 195–220.

Hudson, J. D., and Jenkyns, H. C., 1969. Conglomerates in the Adnet Limestones of Adnet (Austria) and the origin of 'Scheck'. *N. Jb. Geol. Palaeont. Mh.*, Jg. H.9, pp. 552–8.

Other references cited in this chapter:

Arrhenius, G., 1963. Pelagic sediments. In: *The Sea*, vol. III. (Ed. M. N. Hill) Interscience, N.Y. pp. 655–727.

Audley-Charles, M. G., 1965. A geochemical study of Cretaceous ferro-manganiferous sedimentary rocks from Timor. *Geochem. Cosmochim. Acta*, **29**, pp. 1153–73.

——, 1966. Mesozoic palaeogeography of Australasia. *Palaeogeography, Palaeoclimatology, Palaeoecology*, **2**, pp. 1–25.

Berner, R. A., 1965. Activity coefficients of bicarbonate, carbonate and calcium ions in sea water. *Geochem. Cosmochim. Acta*, **29**, pp. 947–65.

Degens, E. T., and Ross, D. A. (Eds.), 1969. *Hot Brines and Recent Heavy Metal Deposits in the Red Sea*. Springer-Verlag, Berlin. 600 p.

Fischer, A. G., and Garrison, R. E., 1967. Carbonate lithifaction on the sea floor. *J. geol.*, **75**, pp. 488–96.

Friedman, G. M., 1965. Occurrence and stability relationships of aragonite, high-magnesium calcite and low-manganesium calcite under deep sea conditions. *Bull. Geol. Soc. Am.*, **76**, pp. 1191–5.

Goldring, R., 1962. The Bathyal lull: Upper Devonian and Lower Carboniferous sedimentation in the Variscan geosyncline. In: *Some Aspects of the Variscan Fold Belt.* (Ed. K. Coe) Exeter Univ. Press. pp. 75–91.

Hudson, J. D., 1967. Speculations on the depth relations of calcium carbonate solution in recent and ancient seas. In: Depth Indicators in Marine Sedimentary Environments. (Ed. A. Hallam) *Marine Geology*, Sp. Issue, **5**, No. 5/6, pp. 473–80.

Price, N. B., 1967. Some geochemical observations on manganese–iron oxide nodules from different depth environments. In: Depth Indicators in Marine Sedimentary Environments. (Ed. A. Hallam) *Marine Geology*, Sp. Issue, **5**, No. 5/6, pp. 511–38.

Riedel, W. R., 1963. The preserved record: palaeontology of pelagic sediments. In: *The Sea*, vol. III. (Ed. M. N. Hill) Interscience, N.Y. pp. 866–87.

Shawe, D. R., Poole, F. G., and Brobst, D. A., 1969. Newly discovered bedded Barite deposits in East Northumberland Canyon, Nye County, Nevada. *Econ. Geol.*, **64**, pp. 245–54.

Shepard, F. P., 1963. *Submarine Geology*. Harper & Row, N.Y. 557 p.

CHAPTER TWELVE

CONCLUSIONS

SEDIMENTARY MODELS, MYTHICAL AND MATHEMATICAL

It should be apparent from the preceding chapters that environmental analysis is at best an imprecise art, rather than a deterministic scientific discipline. This is due to the diverse processes which control a depositional environment and hence the extremely variable nature of the resultant sedimentary facies.

One can never actually prove the depositional environment of a rock as one can derive a mathematical formula or repeat a physical or chemical experiment. We can describe the facies (effect, result), we can only guess at the environment (cause, process).

There are basically two possible philosophic approaches to the environmental analysis of sediments and the utilization of this knowledge. Consider the following simple problem: two boreholes, a few miles apart, pierced a time-equivalent carbonate formation. In one well the rock is an interbedded microcrystalline dolomite : pellet limestone facies. The corresponding interval in the second well is an argillaceous calcilutite with small pelagic foraminifera. Question: what is the nature of the transition between these two facies?

The first solution to this problem might use the following line of reasoning: the dolomite : pellet limestone facies is comparable to the deposits of Recent sabkhas and lagoons and the calcilutite facies is analogous to Recent open marine sediments forming below wave base. Between these two environments it is common today to find a high-energy environment of skeletal or oolite sand shoals; alternatively a reef might be present. The occurrence of a similar facies is predicted between the two wells in question.

The second way of solving this particular problem might run like this: of 100 analogous situations on record, 90 have a reef or calcarenite facies separating the two already discovered. There is, therefore, a probability of 0·90 that the same will hold true for this particular case.

Both approaches may produce the right answer, but while the first is deductive, based on subjective understanding of Recent environments and deposits, the second is empirical and based on a mathematical analysis of objective criteria. The first approach is typical of geologists (as this book illustrates), while the second is more appropriate for a computer.

Both the geologist and the computer need some kind of conceptual framework to solve the problem. Thus the geologist has to have some idea of what Recent sabkhas, lagoons, shoals, and marine platforms are like. These mental pictures will be a function of his field experience, of the books and papers he has read, and of the prejudices and will-power of his teachers. The computer, on the other hand, will have no 'picture' of the environments, but a programmer will have previously fed it data which objectively classify all sedimentary facies and a large number of case histories. Even though the computer does not need to 'understand' recent processes to manipulate facies (results), both computer and geologist must have a logical basis for structuring and classifying the data. The data in the computer must have been analysed and programmed by a geologist liable to the same subjective prejudices as his less sophisticated counterpart.

In both cases, therefore, though the geologist has to interpret the data, which the computer does not, both have to have a conceptual framework of sedimentary facies.

As discussed in the first chapter of this book (p. 2), many writers have considered the problem of classifying and defining sedimentary environments and the facies which they generate. This approach has led to the concept of the sedimentary model. This states that there are, and always have been, a number of sedimentary environments which deposit characteristic facies. These may be classified into various ideal sedimentary systems or models. This concept is fundamental to this book. The case histories in it have been selected as examples which most closely approximate to the writer's subjectively defined ideal models.

There is, however, no such thing in nature as an ideal or typical fluviatile environment or facies, just as the concept of the ideal cabbage exists only in the minds of judges of horticultural shows.

Attempts have been made, though, to define environments and facies mathematically. These projects are now becoming more and more feasible, using the abilities of computers to store and statistically manipulate large amounts of data.

An early attempt to define a sedimentary system mathematically was made by Sloss (1962) and amplified by Allen (1964). The formula proposed by Sloss is:

$$\text{Shape} = f(Q, R, D, M)$$

where Q is the volume of the detritus supplied per unit time; R is the rate of subsidence; D is a dispersal factor of the rate of transport; M is the composition of the material.

This is essentially a process : response model which expresses both the dynamics of the environment as well as the end result. Since we can only guess at the process, but observe its result, it may be safer, to begin with, to define a facies by its five parameters thus:

$$\text{Facies} = \Sigma(G, L, Ss, Pp, F)$$

where G is geometry, L is lithology, Ss is sedimentary structures, Pp is the palaeocurrent pattern, and F is its palaeontology.

Both these formulae are insoluble until it is possible to numerate the parameters (how do you express a palaeocurrent pattern mathematically in a way which can be integrated with a mathematical statement of the associated fossils?).

This approach is, however, a useful starting point to consider ways of defining facies and diagnosing environments mathematically.

Consider first a very simple case. Ignoring geometry and palaeocurrents for the moment, imagine that a facies can be defined only by its chemical composition, grain size, and fauna, i.e.

$$\text{Facies} = \Sigma(C, G, F)$$

where C is the chemical composition, G is the depositional grain size, and F is its palaeontology. Assume too that these three parameters can be calibrated on linear scales. Chemical composition ranging, say, from silica, through calcium carbonate to evaporite minerals; grain size ranging from conglomerate to clay, and fossils being interpreted on a scale which reads from marine through brackish and freshwater to terrestrial.

On this basis it is possible to define a given facies in three-dimensional space using these three variables as mutually perpendicular vector axes (Fig. 12.1).

This is an extremely naïve way of structuring a very complex problem, but it points the way to a more sophisticated approach. A facies is the sum of many more variables than can be expressed graphically in three-dimensional space. It is, however, possible to position objects defined by many variables in multi-dimensional space using the statistical technique of factor analysis (Harbaugh and Merriam, 1968, p. 179). The human mind is incapable of envisaging more than four dimensions (the fourth, time, is not applicable to this situation), but this can be done by a computer.

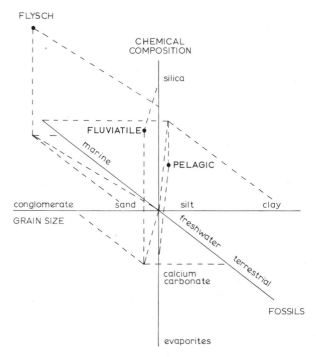

Figure 12.1 Facies fixed in three-dimensional space using their chemistry, grain size and palaeontology as vectors.

If one can numerate all the various parameters of a facies, it is then possible to use factor analysis to fix its position in multidimensional space. This would prove or disprove the concept of the sedimentary model. If this idea is valid, one would predict clusters in space of fluviatile facies, reefs, and so on. These clusters should be composed of both Recent and ancient case histories. Each cluster would represent an 'ideal' model, though a hole in the centre of a cluster would mean that a 'type' specimen was yet to be found (if it exists). Isolated

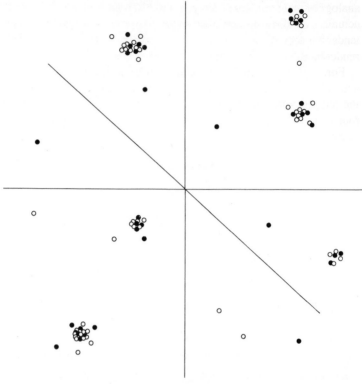

Figure 12.2. Three-dimensional representation for facies fixed in multi-dimensional space using factor analysis. Black spheres indicate location of ancient facies, white spheres indicate Recent environment case histories. Clusters of spheres indicate location of various sedimentary models (e.g. fluviatile, reef, etc.). Randomly distributed spheres indicate case histories transitional between commonly occurring types.

case histories scattered through space between clusters would represent transitional environments (Fig. 12.2).

It must be admitted that this approach to sediments is still highly speculative. It is, however, an interesting way of looking at rocks and, if it can be made to work, will have a major impact on sedimentological studies. The main effect of such a technique would be that facies could be objectively defined and compared with one another. The closest match, or model for a new case history could be calculated and the result used to predict the location of economic materials in the facies under examination by comparison with its analogues. At no stage in the operation would it be necessary actually to diagnose the environment of a facies. The kind of tendentious subjective arguments used in this book would be rendered obsolete.

For the present, however, sedimentary models remain ill-defined and subjective. So long as its limitations are realized, the concept of the sedimentary model is useful both in teaching and as a predictive tool in industry (e.g. Moore, 1967). Some kind of conceptual model is essential for any imaginative kind of interpretation (Fig. 12.3).

Figure 12.3. Some kind of conceptual model is essential for any imaginative kind of interpretation. Anon. 1967. Reproduced from *Atlas* 1, No. 4, p. 62, by courtesy of Elsevier Publishing Company.

Accordingly, this book concludes with an attempt to define some of the various facies models which can be recognized. These are a statement of the subjective opinions of this writer, and should be regarded as guidelines rather than articles of faith.

REFERENCES

Allen, P., 1964. Sedimentological models. *J. Sediment. Petrol.*, **34**, pp. 289–93.

Harbaugh, J. W., and Merriam, D. F., 1968. *Computer Applications in Stratigraphic Analysis*. J. Wiley, N.Y. 282 p.

Moore, P. G., 1967. The Use of Geological Models in Prospecting for Stratigraphic Traps. *Proc. Seventh World Petrol. Cong.*, **2**, Elsevier, Amsterdam, pp. 481–6.

Sloss, L. L., 1962. Stratigraphic models in exploration. *J. Sediment. Petrol.*, **32**, pp. 415–22.

Tables attempting to summarize the diagnostic features of the major sedimentary environments

FLUVIATILE 1

Alluvium of braided rivers

Geometry: The facies as a whole may be prismatic, fan-shaped, or blanket-shaped. Internally it will consist of a down slope trending complex of channel shoestrings of various grades of sediment.

Lithology: Predominantly coarse-grained clastics with conglomerates and coarse sandstones. Rare finer sands and silts. Often red-coloured.

Sedimentary structures: Large channels of cross-bedded and flat-bedded sandstones with occasional quicksand structures. Rare conglomerate-floored abandoned channels infilled by laminated silts with thin rippled sand units and desiccation cracks.

Palaeocurrents: Very low scatter of data at individual sample stations. Regionally may describe fan-shaped arcs.

Fossils: Very rare abraded disarticulated bones of terrestrial vertebrates and plant debris.

FLUVIATILE 2

Alluvium of meandering rivers

Geometry: Sheet; may be composed of alternating sand and shale beds in units of about 5–10 ft. Sand units may be composed of coalesced channel complexes or occur as discrete shoestrings enclosed in shale.

Lithology: Sands and shales present in about equal proportions or sand subordinate. Some conglomerates and coals. Sand of coarse to fine grain. Occasionally predominantly red-coloured.

Sedimentary structures: Tendency for beds to be arranged in fining-upward sequences of grain size with the following structures: scoured and channelled erosion surface overlain by massive or rudely bedded conglomerate; overlain by cross-bedded, flat-bedded, and rippled sands which grade up into siltstones. These are laminated and contain desiccation cracks and thin laminated and rippled fine sand units.

Palaeocurrents: Unimodal at a point with a wide scatter of readings, regionally arranged parallel to the palaeoslope.

Fossils: Channel sands may contain plant debris, rolled disarticulated bones of terrestrial animals and freshwater aquatic vertebrates and shellfish. Overbank floodplain silts may contain plant debris, rootlet horizons, and coal beds.

EOLIAN

Geometry: Irregular sheet.

Lithology: Typically clean, well-sorted medium to fine sandstones. Occasionally gypsum, shell-sand, or silt (loess).

Sedimentary structures: Large scale cross-bedding with set heights of up to 100 ft. Flat-bedding and rare penecontemporaneous deformation may be present. Widely spaced low amplitude ripples.

Palaeocurrents: Dip directions of cross-bedding may be very variable at outcrop due to complex dune morphology. Regional trend of palaeocurrents may be present but are not slope controlled.

Fossils: Generally absent. Rare footprints and plant debris. Coastal dunes may contain, and be exclusively composed of, abraded marine fossils.

LAKE DEPOSITS

Geometry: Irregular sheet.

Lithology: Predominantly fine-grained, argillaceous, calcareous, or evaporitic. Coarse marginal clastics may be present.

Sedimentary structures: Lamination, varves, turbidites, ripples, and desiccation cracks. Cross-bedding and channelling in marginal facies.

Palaeocurrents: Mainly unidirectional and centripetal to the deepest part of the lake. Along the shore bipolar palaeocurrents may indicate onshore and offshore flow. Wave-generated currents may move in any direction regardless of bottom slope.

Fossils: Freshwater, often diverse and very well preserved. Land plants and animals may be washed in. Fine-grained marginal facies may contain coals.

LOBATE SHORELINES (DELTAS)

Geometry: Prismatic or fan-shaped.

Lithology: Clay, silt, and sand present in varying proportions with a tendency to be arranged in upward-coarsening sequences.

Sedimentary structures: Shales in the lower part of a sequence (prodelta) laminated, rarely rippled. Central part of the sequence may have fans or channels filled with turbidite sands. Slumps and slides may be present. Overlying delta platform consists of a radiating complex of cross-bedded channel sands with intervening interlaminated, laminated, ripple, and bioturbated clay, silt, and very fine sand.

Palaeocurrents: Unimodal, may be regionally radiating both in the slope turbidites and channel sands.

Fossils: Marine fauna in lower fine-grained part of sequences; brackish and freshwater fossils present in upper part together with plant debris, rootlet beds, and coals.

CLASTIC LINEAR SHORELINES
(BARRIER ISLAND AND SHOAL COMPLEXES)

These consist of four vertically or laterally juxtaposed facies: alluvial, lagoonal and tidal flat, barrier sand, and open marine. The first of these has been described already, the last is described in the next section. The two intermediate facies will now be described separately.

Lagoon and tidal flat complex

Geometry: Sheets or shoestrings parallel to the palaeostrike.

Lithology: Predominantly clays, silts, and fine sands.

Sedimentary structures: Laminated, rippled, and bioturbated. Rarely with desiccation cracks and conglomerate-lined channels infilled with obliquely inclined shale due to tidal gullies.

Palaeocurrents: Extremely variable, little to measure.

Fossils: Vertebrates and invertebrates ranging from marine, through brackish to freshwater. Shell reefs (especially of oysters) particularly characteristic.

Barrier island complex

Geometry: Sheets or shoestrings parallel to the palaeostrike.

Lithology: Coarse, medium, fine, and very fine sands. Tendency to be matrix-free, and well-sorted. Either terrigenous (predominantly quartzose) or calcarenitic.

Sedimentary structures: For a regressive barrier the following sequence of structures and grain size may be present: transitional base with laminated argillaceous open marine (sub-wave base) facies, grading up into interlaminated, rippled, and burrowed argillaceous silt and very fine sand. This in turn passes up into fine and medium sands with regular parallel stratification, horizontal or gently seaward dipping. Rare troughs and tabular planar cross-beds. Occasional channels (scoured by tidal currents surging in and out through gaps in the barrier). Abrupt upper contact with lagoonal facies. For a transgressive barrier the sequence previously described is reversed.

Palaeocurrents: Cross-beds dip predominantly onshore, sometimes bipolar. Bipolar pattern typical of tidal channels.

Fossils: Bioturbation well developed in transitional zone between barrier sand and open marine facies. Sand itself contains, and may be largely composed of, shallow marine benthonic fossils.

CARBONATE SHORELINES AND SHELVES

Three sub-facies may be defined as follows, from sea to land:

X Zone: open marine sub-wave base: Shales, argillaceous and siliceous limestones, thinly bedded and laminated. Intraformational conglomerate-covered penecontemporaneous erosion surfaces may be present ('hardgrounds'). Well-preserved fauna of pelagic and, rarer, benthonic organisms. Algae absent.

Y Zone: high-energy shoal and barrier complex: Shell and oolite calcarenites, occasionally reefal. Cross-bedded and flat-bedded. Containing, and composed of fragments of, marine shellfish, algae, bryozoa, corals, and larger benthonic foraminifera.

Z Zone: lagoon and sabkha: Faecal pellet limestones, occasionally laminated, desiccation cracked, and with intraformational conglomerates. Often burrowed, with sparse well-preserved, marine fauna. Interbedded with, and pass shoreward into, microcrystalline dolomite with laminae and nodules of anhydrite (may be replaced by gypsum or calcite) and other evaporite minerals.

REEF

Geometry: In plan linear, sub-circular, or, rarely, atoll-shaped.

Lithology: Three lithofacies generally recognizable: (*i*) calcilutites, calcarenites, and pellet limestones (back-reef lagoon); (*ii*) biolithite: *in situ* skeletons of calcium carbonate secreting organisms, may be completely recrystallized, often dolomitized, obliterating original fabric (reef core); (*iii*) skeletal calcarenites and calcirudites with micrite matrices (reef talus). Back-reef facies may grade into sabkha type evaporites. Reef front may be laterally equivalent to 'basinal' evaporites.

Sedimentary structures: Back-reef facies: laminated, rarely bioturbated and with desiccation cracks. Reef core: massive. Reef talus:

poorly developed bedding dipping off reef front; slides and slumps. May grade basinward into carbonate turbidites.

Palaeocurrents: Seldom anything to measure. Slumps, slides, and turbidites may indicate palaeoslope often of only local (reef-generated) significance.

Fossils: Reef core (if not extensively recrystallized) characterized by an abundant fauna with large populations of many species. Reef framework formed of calcareous algae, stromatoporoids, bryozoa, and corals, associated with many other sessile and mobile organisms.

FLYSCH
(*marine ? turbidite deposits*)

Geometry: Sheets, fans, or channels.

Lithology: Alternations of sand and shale. Sands very variable in composition ranging from skeletal shell sands, through proto-quartzites to greywackes. The latter is probably most typical especially in older (pre-Tertiary) examples.

Sedimentary structures: Typical bedding aspect shows monotonous alternations of sands and shales. Sand units seldom greater than 10 ft, generally less than 1 ft. Bases of sands show diverse suite of erosional structures; graded internally with a tendency for the following sequence of structures to be present from base to top: massive, laminated, micro-crosslaminated (with or without convolute lamination), laminated transition to shale. Slides, slumps, and channels may also be present.

Fossils: Shales may contain pelagic fossils, possibly of deep water aspect. Sands contain abraded shallow water benthonic fossils and, sometimes, plant debris.

PELAGIC

Geometry: Often hard to determine since this facies is typical of tectonically disturbed mountain chains. Typically occurs in thin sequences above shelf carbonates and below flysch facies.

Lithology: Calcilutites, sometimes red and nodular; red clays, highly ferruginous and manganese rich, radiolarian cherts.

Sedimentary structures: Thinly bedded and laminated. Occasionally with ripples and 'hardground' horizons.

Palaeocurrents: Seldom anything to measure.

Fossils: Characterized by an absence of benthos, though some burrowing may be present. Typically with a sparse but well-preserved pelagic fauna of radiolaria, foraminifera, thin-shelled molluscs, and marine vertebrates. Tendency for fossils to show penecontemporaneous corrosion and preservation of shells of original calcitic composition, over those originally aragonitic.

INDEXES

AUTHOR INDEX

SUBJECT INDEX